Land and Heritage:
The Public Interest in Personal Ownership

First published in April 1982

© THE INSTITUTE OF ECONOMIC AFFAIRS 1982

All rights reserved

ISSN 0073-2818
ISBN 0-255 36151-3

Printed in England by
GORON PRO-PRINT CO LTD
6 Marlborough Road, Churchill Industrial Estate, Lancing, W. Sussex

Text set in 'Monotype' Baskerville

CONTENTS

	page
PREFACE	7
THE AUTHOR	9
GLOSSARY	10
INTRODUCTION	15
I OWNERSHIP AND WEALTH	17
The ownership dimension	17
Saving and consumption	18
(i) Intertemporal shift in consumption	19
(ii) The power to consume	20
(iii) Wealth without consumption	22
A difference of opinion	23
Saving in perpetuity	24
The categories of saving and wealth	25
Anomalies and inconsistencies	27
Policy towards surplus value	29
Other implications for policy	30
II IS LAND UNIQUE?	32
Doctrines of differentiation	32
Adam Smith and the Physiocrats	32
Ricardo versus Smith	34
The Physiocrats and Henry George	35
Assimilation of land to other assets	35
Land not fixed in quantity	37
Elements of difference	38
Objective balance and subjective imbalance	38
Consequences of government action	41

III Systems of Land Ownership — 46

- The concept of capital-efficiency (maximising the value of ownership) — 46
- Criteria of an economic and fiscal system which maximises the value of ownership — 46
- The creation of property rights as a solution to economic problems — 48
- Market and state: a schematic classification of landholding systems — 50
  1. Private property: single ownership — 50
  2. Private property: communal ownership — 51
  3. State property — 52
  4. Joint ventures — 53
  5. State control of private property — 53
- Northfield and CLA classifications — 54
- The income-efficiency of landholding systems — 55
- The capital-efficiency of landholding systems — 58
  - The four criteria of capital-efficiency — 60
- Land reform — 63

IV Effects on Third Parties — 66

- Principles governing the public interest in landholding systems — 66
- Social responsibilities — 68
- Interests of employees — 70
- Interests of tenant farmers and new entrants to farming — 70
- Interests of residential tenants — 71
- Interests of neighbours — 72
- Recreational interests — 73
- Interests of voters — 74
- A survey of public opinion — 75

V The Countryside and the Heritage — 77

- Real and factitious difficulties of conservation — 77
- Friends and enemies of conservation — 77
- Vocational ownership — 80
  - The four values of an estate — 81

Summary of Hobart Paper 93

Land and Heritage: The Public Interest in Personal Ownership
BARRY BRACEWELL-MILNES

1. Land is not only a *personal* asset, being worth more to the owner than to strangers; it is also *proprietary wealth*, being worth more to society when owned by private persons than by the state or its agencies.

2. As both a personal asset and proprietary wealth, land is particularly unsuitable for government interference; in state hands property rights become *sterile*.

3. The interests of third parties in land—employees, tenant farmers, potential new entrants to farming, residential tenants, neighbours, recreation-seekers—are generally best served by the larger personal landowner.

4. There is a strong element of vocation in the ownership of heritage assets—in town as well as country—especially if they are inherited. The public interest in their personal ownership may be much larger than the value of the same assets to the owners in their private capacities.

5. Personal ownership can help heritage estates and rural communities survive as living entities.

6. Britain is squandering a precious and irreplaceable resource through a tax régime which discourages and frustrates the ownership of heritage assets by their historic owners.

7. The valid case for abolishing capital gains and capital transfer taxes for all assets—or at least for reducing them to rates which can be paid out of income—is especially strong for heritage assets.

8. If capital taxes are retained for heritage assets, they should be levied on something near the private value in current use and not on the much higher break-up value.

9. Public access is not a prerequisite of the public interest in personally-owned heritage assets. But if it is made the condition for reduced taxation, it should be satisfied by occasional access or access by appointment.

10. Reforms along these lines would diminish the demands on government funds for the upkeep of heritage assets.

Hobart Paper 93 is published (price £3.00) by

THE INSTITUTE OF ECONOMIC AFFAIRS
2 Lord North Street, Westminster
London SW1P 3LB   Telephone: 01-799 3745

IEA PUBLICATIONS

Subscription Service

An annual subscription is the most convenient way to obtain our publications. Every title we produce in all our regular series will be sent to you immediately on publication and without further charge representing a substantial saving.

Subscription rates*

Britain: £15·00 p.a. including postage.
£14·00 p.a. if paid by Banker's Order.
£10·00 p.a. teachers and students who pay *personally*.

Europe and South America: £20 or equivalent.

Other countries: Rates on application. In most countries subscriptions are handled by local agents.

*These rates are *not* available to companies or to institutions.

To: The Treasurer, Institute of Economic Affairs,
    2 Lord North Street,
    Westminster, London SW1P 3LB.

I should like to subscribe beginning.....................................
I enclose a cheque/postal order for:

☐ £15·00

☐ Please send me a Banker's Order form

☐ Please send me an Invoice

☐ £10·00 [I am a teacher/student at..............................]

Name..

Address...

....

Signed... Date...................

HP93

The public interest in vocational ownership	84
Extent and direction of the public interest in vocational ownership	85
Ingredients and evidence	86
High cost of state ownership and maintenance	88
Qualifications	91
Conservation bodies and pressure groups: rôle of third-party interests	91
A going concern	94
Unvisited, unblest?	96
Conclusion	98
VI THE URBAN LANDLORD	100
Rus in urbe	100
Rent control and 'leasehold enfranchisement'	100
The urban estate	102
Planning strategy and conservation	103
Conservation and planning details	105
Conclusion	106
VII THE BURDEN OF TAXATION	107
The weight of the burden	107
A tax on vocation	107
Two senses of 'taxable capacity'	108
The taxation of saving in perpetuity	110
Tax alleviation or tax reduction?	111
Recommendations for public policy	113
VIII CONCLUSIONS	115
QUESTIONS FOR DISCUSSION	117
FURTHER READING	118
SUMMARY	*Back cover*

PREFACE

The *Hobart Papers* are intended to contribute a stream of authoritative, independent and lucid analyses to the understanding and application of economics to private and government activity. The characteristic theme has been the optimum use of scarce resources and the extent to which it can best be achieved in markets within an appropriate framework of law and institutions or, where markets cannot work, in other ways. Since in the real world the alternative to the market is the state, and both are imperfect, the choice between them effectively turns on judgement of the comparative consequences of 'market failure' and 'government failure'.

Can land be fitted into this framework of economic analysis? Is it in economically significant ways different from other economic assets or factors of production? In Hobart Paper 93, Dr Barry Bracewell-Milnes, an economist well versed in classical economics but with an independent mind that leads him to unconventional analyses and unorthodox conclusions, provides a refreshingly stimulating study of the economics of land ownership. Some economists will differ from him, but his *Paper* points to aspects of the subject and raises questions that might stimulate reconsideration of established approaches.

Dr Bracewell-Milnes reviews the orthodox interpretation of land ownership, and finds it wanting. The pioneering flavour of his approach leads him to a closely reasoned discussion that requires correspondingly close attention to appreciate its significance. His independence of mind was demonstrated in 1973 when on grounds of principle he chose dismissal rather than resignation from his position as Economic Director of the Confederation of British Industry; his support for market-oriented rather than corporatist policies has been largely vindicated by subsequent developments.

In the course of his analysis Dr Bracewell-Milnes makes some intriguing statements and reaches provocative conclusions: the personal ownership of land gives the owner a sense of satisfaction that does not always accompany the ownership of other assets; land as heritage is worth more to society when owned privately than collectively; the satisfaction of property rights in land becomes sterile when it is owned collectively, and so on. And the analysis leads the author to conclusions for public policy on taxation.

Whatever the argument on the approach that the personal ownership of land gives unique satisfaction both to the owners and to society in general, it is a direct challenge to the long-established view that the public interest was served only by the collective ownership of land through nationalisation. There was always a large, often unstated, assumption that 'publicly-owned' land would be administered in the interest of the public. Perhaps experience of other forms of nationalisation has taught the lesson that, as in the railways, the mines, the schools and the hospitals, assets that are not appropriated to individuals are not husbanded because no individual can identify his interest. The reply to the argument for nationalisation was that only individual ownership would ensure individual care, protection and conservation. Dr Bracewell-Milnes reinforces the objection to 'public ownership' by a refined statement of the positive advantages both for the owner and for society.

For many years the words 'private' and 'public' have been misused or misunderstood. 'Private' has been used to imply consequences contrary to the interest of the public. 'Public' has been used to imply consequences necessarily favourable to the interest of the public. Post-war experience has confirmed the views of the classical economists, going back to David Hume and earlier, that the best interests of the public are served by private ownership. This *Paper* argues that the ownership should be not only private but personal. It could therefore, by extension, be read as a criticism not only of public ownership but also of non-personal ownership in large private organisations. Recent discussion on methods of re-creating personal interest in large organisations is a facet of the economics of personal ownership.

The Institute has to thank Arthur Shenfield and other economists for reading early drafts and making observations that have been taken into account by the author in his final revisions. The constitution of the Institute requires it to dissociate its Trustees, Directors and Advisers from the analyses and conclusions of its authors, but it offers this *Hobart Paper* as a stimulating re-appraisal of a central component of economic life in modern society.

March 1982

ARTHUR SELDON
Advisory Director

THE AUTHOR

DR BARRY BRACEWELL-MILNES was born in 1931 and educated at Uppingham, at New College, Oxford, where he read Classical Moderations and then changed to Economics, and at King's College, Cambridge, where he took his doctorate.

Dr Bracewell-Milnes now works as a consultant to academic and industrial bodies on government and international fiscal and economic policy. He was Economic Director of the Confederation of British Industry, 1968-73. Since leaving the CBI he has been Economic Adviser to The Institute of Directors, and his other appointments during the period have included Economic Consultant to the Fiscal-Economic Institute, Erasmus University, Rotterdam.

He is the author of over a dozen books on taxation and other economic subjects, including *The Measurement of Fiscal Policy: An Analysis of Tax Systems in Terms of the Political Distinction between 'Right' and 'Left'* (1971), and *Is Capital Taxation Fair? The Tradition and the Truth* (1974). His books on tax avoidance and evasion include *Tax Avoidance and Evasion: The Individual and Society* (Panopticum Press, Upminster, 1979, through The Alternative Bookshop, Covent Garden, London WC2); *International Tax Avoidance: Volume A, General Report* and *Volume B, Country Reports* (Kluwer, Deventer, 1979); and *The Economics of International Tax Avoidance: Political Power versus Economic Law* (Kluwer, 1980).

The argument of the present *Hobart Paper* is extended from land and heritage assets to assets of all kinds in his latest book, *The Taxation of Industry: Fiscal Barriers to the Creation of Wealth* (Panopticum Press, 1981).

For the IEA Dr Bracewell-Milnes has written 'The Economics of Tax Reduction', in *Taxation: A Radical Approach* (IEA Readings No. 4, 1970); 'Market Control over Land-Use "Planning"', in *Government and the Land* (IEA Readings No. 13, 1974); 'The Fisc and the Fugitive', in *The State of Taxation* (IEA Readings No. 16, 1977); and an Epilogue, 'Is Tax Avoidance/Evasion a Burden on Other Taxpayers?', in *Tax Avoision* (IEA Readings No. 22, 1979).

GLOSSARY

BREAK-UP VALUE—The break-up value of an estate, like that of a company, is the sale value of its component parts. This is less than the *social value in current use* if there is a public interest in the preservation of the estate as a going concern. See also *personal value in current use* and *private value in current use*.

CAPITAL COST OF VOCATION—Excess of *personal value in current use* over *private value in current use*.

CAPITAL-EFFICIENCY OF OWNERSHIP—The realisation of the value of ownership as such. The creation of *surplus value* through the proprietary ownership of proprietary assets and the personal ownership of personal assets. See *income-efficiency of ownership*.

COMMITMENT—The value of emotional capital sunk in *vocational ownership*.

COMMUNAL WEALTH—Assets worth more to society when owned by the state than when owned by private persons.

CONCEPTUAL OVER-VALUATION OF ASSETS FOR PURPOSES OF CAPITAL TAXATION—Systematic over-valuation of assets at a market value largely determined by taxpayers subject to capital transfer tax at lower rates or (like companies) not subject to it at all, without allowance for the present value of future liabilities to capital transfer tax. Heritage assets are systematically over-valued when *break-up value* is in excess (often far in excess) of *private value in current use*.

CREATION OF PROPERTY RIGHTS—The development of law or custom making private rights out of rights which were formerly annihilated by *dilution* (or anarchy). See *manufacture of property rights*.

DILUTION (DISPERSAL, DISSIPATION)—Ownership dilution, dispersal or dissipation is the destruction of value (or the prevention of its realisation) through the reduction of ownership *intensity*. Value can be destroyed (or remain unrealised) not only through government ownership of *proprietary* assets and impersonal ownership of *personal* assets but also through the lack of a system of property rights sufficiently sensitive to maximise *income-efficiency* and *capital-efficiency*.

DISPERSAL—See *dilution*.

DISSIPATION—See *dilution*.

DIVISION (FRAGMENTATION)—The division of a capital interest into a large number of parts may, but need not, cause *dilution*; in particular, an interest need not be diminished at all by its division into shares in a company.

EMINENT DOMAIN—Power of compulsory acquisition in what is alleged to be the 'public interest' for what is alleged to be 'just compensation'.

EXTERNAL ILLIQUIDITY—See *objective illiquidity*.

EXTERNAL SOCIAL VALUE IN CURRENT USE—The excess of the social value in current use over the break-up value.

FARMERS' PARADOX—The consequence of capital taxation that a rise in the value of an asset which apparently makes a taxpayer richer makes him in reality poorer if its value to him exceeds the new higher price.

FRAGMENTATION—See *division*.

ILLIQUIDITY—See *objective illiquidity, subjective illiquidity*.

INCOME-EFFICIENCY OF OWNERSHIP—The realisation of the value of ownership in terms of discounted cash flow. See *capital-efficiency of ownership*.

INCOME EQUIVALENT OF CAPITAL TRANSFER TAX—The cost of paying for capital transfer tax out of income.

INTENSITY—Ownership intensity is the creation or preservation of value through an articulated system of property rights, which promotes *capital-efficiency* as well as *income-efficiency*. See *dilution*.

INTERNAL ILLIQUIDITY—See *subjective illiquidity*.

IRREPLACEABILITY—Assets are irreplaceable if their supply cannot be increased by incentives. Examples are minerals in the ground, works of Old Masters, and a family's commitment to keeping up its home.

LIQUIDITY—See *marketability, objective illiquidity, subjective illiquidity*.

Manufacture of Property Rights—Transfer of genuine or creation of factitious property rights through government interference. The economic loss to society normally exceeds the gain. See *creation of property rights*.

Marketability—Assets are marketable in so far as they are not illiquid either objectively or subjectively.

Objective Balance—The arithmetical identity between debts and credits or between the corresponding flows on income account.

Objective or External Illiquidity—The excess of the long-term value of an asset over its short-term value at any given rate of interest. See *marketability*.

Ownership Dilution, Dispersal or Dissipation—See *dilution*.

Ownership Intensity—See *intensity*.

Partition of Property Rights—The different elements of property rights may be *apportioned* between different citizens, each separate element remaining undiluted. See *division*.

Personal Asset—An asset worth more to its owner than to strangers by reason of family or other personal links. A personal asset is also by definition a *proprietary asset*, though not all proprietary assets are personal assets.

Personal Saving—Assets held by an individual directly and not through an intermediary or a chain of intermediaries.

Personal Value in Current Use—The value of an estate to its owner as a going concern including the *surplus value* of *personal assets*. Where ownership is *vocational*, personal value in current use exceeds *private value in current use*, and the wealth constituting this excess is immaterial and cannot be realised. *Break-up value* comes between personal and private value in current use, and the same relationship can hold good for assets in which there is little or no public interest, such as a family farm.

Private Value in Current Use—The value of an estate to its owner as a going concern excluding the *surplus value* of *personal assets*. Where ownership is *vocational*, private value in current use falls short of *break-up value*.

Proprietary Asset or Wealth—An asset worth more to society when owned by private persons than by the state or its agencies. See *personal asset*.

PROPRIETARY COMMUNITY—A community (such as a hotel) in which the interests of the various users are reconciled by their common relationship with the proprietor.

PROPRIETARY LAND UNIT—An area of land managed as a single entity and co-extensive in its physical dimensions with vested rights of user, disposition and alienation.

PROPRIETARY PRINCIPLE OF WEALTH CREATION—The principle that wealth is created by the transfer of *proprietary assets* from the hands of the state or its agencies into the hands of private persons. It applies to all *personal assets*, since a personal asset is also by definition a proprietary asset; and additional wealth is created if personal assets are owned by those with family or other personal links rather than by strangers.

PROPRIETARY WEALTH—See *proprietary asset*.

PUBLIC GOOD—A good or service of which the use by one citizen does not diminish the supply available to another.

QUALITATIVE BURDEN—See *quantitative burden*. The qualitative burden is the ratio of tax revenue to some measure of national income *after* allowance for differences in the painfulness and perceptibility of different taxes.

QUANTITATIVE BURDEN. See *qualitative burden*. The quantitative burden is the ratio of tax revenue to some measure of national income *before* allowance for differences in the painfulness and perceptibility of different taxes.

SAVING IN PERPETUITY (PERMANENT SAVING)—Saving which is never realised, whether by accident or design.

SOCIAL VALUE IN CURRENT USE—Capitalisation of the public interest in an estate as a going concern. The social value in current use exceeds the *personal value in current use* if the estate is cheaper to run as a going concern privately than publicly and if there is a public interest in the maintenance of the estate as a going concern. The social value in current use exceeds the *break-up value* if there is a public interest in the maintenance of the estate as a *personal* (or even *proprietary*) asset (in particular, if a building intended to be used as a family home continues to be used for this purpose). The public interest consists partly of the interest of the owner (the excess of personal value in current use over break-up value) and partly of the additional interests of outsiders (the excess of social value in current use over personal value in current use). The excess of the social value in current use over the *personal value in current use* includes the capitalised value of any *income-efficiency* obtainable through lower costs of running a property in personal rather than government ownership. See *external social value in current use*.

STERILE OWNERSHIP—Destruction of the *surplus value* obtainable through ownership (or the prevention of its realisation) when proprietary assets are owned by the government or personal assets are owned impersonally. See *surplus value*.

SUBJECTIVE OR INTERNAL ILLIQUIDITY—The difficulty or impossibility of realising the market value of assets in personal ownership when the *personal value in current use* exceeds the *private value in current use*. *Objective* or *external illiquidity*, by contrast, is the difficulty or impossibility of selling impersonal assets in the short term at or near the capitalised value of their long-term income yield. See *marketability*.

SUBJECTIVE IMBALANCE—The difference between the importance of debts and of the correlative credits as a result of different subjective assessments by debtors and creditors. Similarly for the corresponding flows on income account.

SURPLUS VALUE—The additional value created by the personal ownership of *personal assets* and the private ownership of *proprietary assets*. See *sterile ownership*.

TAX ON VOCATION—A tax on altruism in earning, saving or owning. See *vocational ownership*.

VACANT-POSSESSION PREMIUM—The difference between the price of vacant-possession land and the price of land sold subject to tenancy, expressed as a percentage of the former.

VOCATIONAL OWNERSHIP—Ownership of an asset of which the *personal value in current use* exceeds the *private value in current use*. See *commitment*.

INTRODUCTION

This *Paper* is about the economics of personal land ownership and how the holding of land by individuals compares with its holding by corporate bodies and government agencies. Who owns the land has been an important political question since early times, and it still is. Our aim is to see what contribution economics can make to answering this question and how far economics can transcend the limitations of a political controversy.

Section I discusses ownership as a dimension of welfare—the creation of wealth through the institution of property or ownership, not only of land, but of other assets as well. This subject has been largely neglected by economists.

Section II considers whether the particular qualities of land affect the argument: can wealth be created more or less easily through the ownership of land than through the ownership of other assets?

Section III suggests a classification and compares ownership by individuals with other forms of possession.

Section IV combines the argument of the three preceding sections to consider how personal land ownership affects outsiders, including the public at large or the rest of the community. These are the *external effects* or *externalities* of economic jargon.

Section V relates the argument to the acknowledged public interest in maintaining and preserving the country's inheritance of art and nature. There is a *public* interest in the *personal* ownership of heritage assets.

Section VI applies the argument to urban estates, abstracting from the separate subject of urban planning and development.

Section VII discusses the present burden of taxation on personal land ownership in Britain and abroad, especially when the value of money is falling substantially from year to year, and goes on to consider the cost it imposes on society as a whole.

Section VIII summarises the conclusions.

I. OWNERSHIP AND WEALTH

The ownership dimension

Utility, the stuff of economic wellbeing in the traditional economic doctrine, is obtainable from ownership as well as from consumption. Ownership represents potential consumption, and the power to consume has value as well as its exercise. The point was made by David Hume in the section of his *Treatise* entitled 'Of property and riches'.[1] Admittedly, Hume starts by saying that 'the distinction, which we sometimes make betwixt a *power* and the *exercise* of it, is entirely frivolous'; but he goes on so to qualify this principle as to reverse it completely in the context that interests us here.

> 'But tho' this be strictly true in a just and *philosophical* way of thinking, 'tis certain it is not *the philosophy* of our passions; but that many things operate upon them by means of the idea and supposition of power, independent of its actual exercise.
>
> The very essence of riches consists in the power of procuring the pleasures and conveniences of life. The very essence of this power consists in the probability of its exercise, and in its causing us to anticipate, by a *true* or *false* reasoning, the real existence of the pleasure.' (Emphasis Hume's.)

All this relates directly to the mainstream of Western economics in which value (and thus welfare) is determined by utility. In the words of W. S. Jevons, the English inventor of modern value theory:

> 'My present purpose is accomplished in . . . assigning a proper place to the pleasures and pains with which the Economist deals. It is the lowest rank of feelings which we here treat. . . . Each labourer, in the absence of other motives, is supposed to devote his energy to the accumulation of wealth . . .
>
> The laws which we are about to trace out are to be conceived as theoretically true of the individual . . . the laws of the aggregate depend of course upon the laws applying to individual cases.'[2]

---
[1] *A Treatise of Human Nature* (1739), Book II, Section X.
[2] W. Stanley Jevons, *The Theory of Political Economy* (1871), 4th Edition, Macmillan, London, 1924, pp. 26-27, 48.

In other words, whatever Hume's 'philosophical' objections to the distinction between a power and its exercise, they do not invalidate the economic distinction between welfare through consumption and welfare through ownership, since economics is rather concerned with 'the philosophy of our passions'; the raw material is the pleasures and pains of individuals, and 'it is the lowest rank of feelings which we here treat'.

Wealth is not valuable only as spending power; it has other advantages and corresponding disadvantages. It may confer political power or standing in society, for example, or the opportunity to perpetuate one's name. Conversely, the ownership of wealth may in some societies diminish a man's standing or impair his career; it often imposes financial and other obligations; and it attracts unwanted attention from burglars, tax gatherers and the like.

Consumption also has secondary qualities. There is a repute of consumption, for example, just as there is a repute of ownership, and the secondary qualities may be more important than the primary—as when a man goes to an opera he knows he will not enjoy merely in order to impress his friends. But consumption differs from ownership in that the primary and secondary purposes of consumption are sharply distinguished even though the secondary purposes may on occasion be the more important. The primary purpose of consumption is to enjoy the goods or services consumed.

Ownership, by contrast, is more complex and has more than one primary purpose. The enjoyment of spending power provides a starting point; but other primary purposes, such as the perpetuation of the family name, may depend on this power's not being exercised.

The complexities of the ownership dimension are the theme of this *Hobart Paper*. How can policy towards the ownership of assets in general and land in particular be so designed as to maximise the economic wellbeing obtainable from the institution of ownership as such?

Saving and consumption

For ownership, as for consumption, there is a distinction between primary and secondary purposes. But the primary purposes of ownership, unlike those of consumption, are various and inter-related and some of them shade into the secondary purposes. The interrelationships of these primary purposes are most conveniently analysed in terms of the relationship between saving and consumption. Is saving intended to increase consumption and, if so, when and in what sense? In what follows, the problem is divided into three components; while in practice they overlap, it is simpler to discuss them separately and there is no difficulty in applying the argument to more complex real situations.

(i) *Intertemporal shift in consumption*
This is the traditional theory of abstinence. The consumer wishes to put money aside, within a year (for a holiday) or a lifetime (for a pension) or just for a rainy day; and the rate of interest makes it attractive for him to do so—i.e., the net-of-tax rate of discount exceeds his subjective rate of discount.[1] The traditional argument ignored inflation, which has become a major pre-occupation only in more recent years. But it can be adapted without difficulty to a world in which prices rise substantially each year: if the real interest rate is to remain positive, the net-of-tax money interest rate must include an element that makes good the fall in the value of the principal caused by the rise in prices.

The traditional theory of abstinence—articulated at a time when extensive freedom from government intervention in the economy made expansion seem normal—assumes a positive (real) rate of interest because the economy is assumed to be expanding. But it can be adapted to situations in which government interference or acts of God make economic decline likely or inevitable. If for either of these reasons the economy is expected to remain in long-term decline, it is rational to accept a negative rate of interest as the price of survival—for example, to do without smoked salmon now in order to have bread and cheese in 20 years' time.

An example of economic decline that owed nothing to errors of government policy is provided by Viking Greenland. The

[1] The percentage rate of discount is $i/(100+i)$, where i is the percentage rate of interest. If the rates of discount and interest are positive, an increase in capital through the re-investment of interest from 100 now to (say) 125 in (say) five years is equivalent to the discounting of 100 in five years to 80 now. If discount and interest rates are negative, the relationships are reversed: 100 invested now yields (say) 80 in (say) five years, and 100 in five years has a present discounted value of 125.

drop in temperature in the later Middle Ages that pushed back the margin of viticulture from Lincoln to the Loire had its effect further north as well. The Viking settlements in Greenland were presented with the choice of turning Eskimo or failing to survive—and they were not willing or able to turn Eskimo.[1] During this period, the rate of interest attributable to abstinence was negative; its function was not to raise the standard of living but to reduce and retard its decline.

The shift in consumption from one period to another increases the present discounted value of consumption to the consumer if the market rate of discount exceeds the consumer's subjective rate. The consumer gains, for example, from postponing consumption if the market rate of discount is 5 per cent and his subjective rate is 4. The same principle holds good if the rates of interest and discount are negative so that the present value of future consumption is higher, not lower, than its future value. The consumer still gains from abstinence if the market rate of discount exceeds the consumer's subjective rate (in other words, if it is a smaller negative)—if the market rate is −4 per cent and the consumer's rate is −5, for example; the volume of future consumption is higher at the market rate than at the consumer's rate even though both are less than the volume of consumption obtainable by spending at once instead of investing at a negative rate of interest.

In order to distinguish intertemporal shifts in consumption from the power to consume, it may be assumed that the savings are invested in an illiquid asset such as an income bond.

(ii) *The power to consume*

Suppose a man has a net-of-tax income of £100, of which he spends £90 and invests the remaining £10 in liquid assets with a net-of-tax yield of 10 per cent. His subjective rate of discount is less than the market rate and so his power to consume in the future is worth more to him now than £10 of immediate spending. In the second year, his income is £101; but saving is attractive to him for the same reason as before, and he spends £90 and saves £11. Similarly, in all subsequent years he spends £90 and saves the sum of the original £10 and the rising income from previous investment.

This is not an intertemporal shift in consumption. The man's consumption is lower in every year than it would be if he were

[1] Gwyn Jones, *The Norse Atlantic Saga*, Oxford University Press, 1964.

saving nothing; there is no rate of discount, positive or negative, at which his consumption is increased. The power to consume is preferred to its exercise. The preference is not irrational or perverse; it is a matter of individual choice.

Intertemporal shifts in consumption are conveniently thought of as shifts within a single human lifetime, whereas the preference for the power to consume over its exercise suggests a time-horizon extending beyond a single generation. These are the simplest and most typical cases, though they are not the only possibilities.

In the example just given, spending remains permanently at £90; the income elasticity of demand for consumption is zero. This logical extreme was assumed in order to simplify the argument. It can be relaxed without weakening the argument, though at the cost of making it more complex.

Suppose that new saving remains at £10 and that spending rises to absorb the increasing income from investments: £90, £91, £92 . . . etc. After 10 years the man's consumption attains the level at which it would have remained if he had saved nothing, and thereafter it exceeds this level. Eventually the missing consumption (£10+£9 . . . . . . +£2+£1) appears to be made good and more than made good. But this is true only at a zero or low rate of interest. It is not true at or anywhere near an interest rate of 10 per cent, at which level the present discounted value of consumption in every year would be increased by saving less and spending more.

The subjective rate of discount, which makes saving attractive to the saver, is below the market rate.[1] The attraction of the power to consume also implies that the discount rate is positive; if the discount rate is negative, the motive for saving is not the power to consume but survival. It follows from the argument in the preceding paragraph that there is a break-even or watershed positive rate of discount below which consumption may be regarded as the purpose of saving and above which it cannot. Thus the four rates of discount are, in descending order:

(i) the market rate;
(ii) the saver's subjective rate;
(iii) the break-even rate;
(iv) zero.

---

[1] This *economic rent* from saving is analogous to consumers' surplus from purchasing.

At the two logical extremes, the saver's subjective rate may coincide with the market rate or with zero; if the subjective rate is zero, so is the break-even rate. In the normal situation where the subjective rate is below the market rate but above zero, the break-even rate is below the subjective rate if the power to consume is the motive for saving. The three bands between the four rates of discount represent:

(i) from market rate to saver's subjective rate, saver's economic rent;
(ii) from saver's subjective rate to break-even rate, power to consume;
(iii) from break-even rate to zero, additional consumption.

The break-even rate is not a purely subjective rate, although it depends on the saver's preferences. If saving is mainly for consumption, the break-even rate is at or near the saver's subjective rate; if it is mainly for the power to consume, the break-even rate is little above zero; and if it is in perpetuity, the break-even rate is zero. Over any finite period during which net saving is positive (i.e., new saving exceeds spending out of old savings), the pattern of positive and negative saving from year to year determines the break-even rate of discount below which aggregate consumption is increased by saving and above which it is reduced. If net saving is nil over the period, the break-even rate and the saver's subjective rate coincide; as net saving increases, the break-even rate falls relatively to the saver's subjective rate. The longer the term of saving in any period, the larger the volume of net saving (since more saving is outstanding at the end of the period, even if it is intended for spending later). So, in any given period, there is a positive relationship between the term of saving and its motivation by the desire for spending power rather than actual spending. If the period is extended into the future without limit, the distinction between short- and long-term saving disappears and the only relevant distinction is between saving which is realised for consumption and saving which is not.

(iii) *Wealth without consumption*

Under the heading 'The power to consume' we have already noticed the difficulty of explaining saving as an activity motivated exclusively by the desire for increased consumption.

Some patterns of saving reduce consumption in every period; others increase it only at rates of discount far below the market rate, the saver's own subjective rate and the rate that would make intertemporal shifts of consumption attractive. The power to consume has a value additional to that of its exercise.

By an extension of the same argument, wealth can have utility even if the connection with consumption is so attenuated as to disappear entirely. Just as at one extreme wealth can have value solely by virtue of increasing consumption through shifts over time, so at the other it can have value directly and in its own right. The power to consume is intermediate between these extremes. Examples of the qualities of wealth that make it desirable in its own right have already been given under 'The ownership dimension'. It serves no purpose to describe these advantages of wealth as yielding psychic consumption; this metaphorical use of the word 'consumption' has nothing to do with its ordinary sense.[1]

A difference of opinion

The consensus among economists has been that the purpose of saving and ownership is to increase consumption. Professor Murray Rothbard, for example, argues:

> 'Saving and consumption are not really symmetrical. All saving is directed toward enjoying more consumption in the future. Otherwise, there would be no point at all in saving. . . . No one wants capital goods for their own sake. They are only the embodiment of an increased consumption in the future. . . . There is nothing, after all, especially sacred about savings; they are simply the road to future consumption'.[2]

This position is fully consistent with the use of saving to achieve intertemporal shifts in consumption. It is partly consistent with the concept of wealth as spending power. But, as we have seen under (ii) above, if the value of the spending power depends on its exercise, the discount rate required to yield an increase in consumption may be so low as to bear no relation to the market rate the saver receives or to the subjective rate he requires. And if the saver re-invests all his addi-

---

[1] These advantages of wealth have economic *value* (they yield psychic *utility* or *satisfaction*); but it is confusing, and even the opposite of the truth, to describe them as yielding psychic *consumption*.

[2] *Power and Market: Government and the Economy*, Institute for Humane Studies, Menlo Park, California, 1970, pp. 74-75.

tional investment income, his consumption falls in every period and there is no rate of discount, positive or negative, at which his aggregate consumption increases.

Henry Simons takes a different line:

> 'The observable fact is that many people save instead of consuming. . . . To assume that all economic behaviour is motivated by desire for consumption goods, present and future, is to introduce a teleology which is both useless and false. . . . In a world where capital accumulation proceeds as it does now, there is something sadly inadequate about the idea of saving as postponed consumption.'[1]

This is fully consistent with (ii) and (iii) above as well as with (i). Simons is right to say that saving may have value independently of the additional consumption it makes possible. Saving that is never spent is saving in perpetuity.

Saving in perpetuity

The concept of permanent saving has been discussed by the author elsewhere.[2] Saving is permanent if the capital is never spent. Consumption of the capital is not merely deferred but permanently forgone. Forgone permanently, not irrevocably: saving does not acquire its quality of permanence at the outset or irrevocably but becomes permanent merely by virtue of being left undrawn forever. The original act of saving is thus continually renewed. The saver enjoys not only the reality of receiving the income the saving generates but also the possibility of drawing down the capital as well. But if the saving is permanent, this possibility is never realised.

The concept of permanent saving is perhaps easiest to understand in the aggregate. Even in a stationary economy the use of capital in the production process requires the existence of a permanent pool of savings if the means of production are privately owned. But there are two separate reasons why saving may be permanent individually as well as in the aggregate.

The first is that the saver may indeed intend his saving to be 'a possession for all time'.[3] *A good man leaveth an inheritance*

---

[1] *Personal Income Taxation*, University of Chicago Press, 1938, pp. 95-97.

[2] *Is Capital Taxation Fair? The Tradition and The Truth*, Institute of Directors, London, 1974, pp. 40, 61-67; 'A Liberal Tax Policy', *British Tax Review*, 2/1976, pp. 115-117.

[3] 'A possession for all time and not merely the exploit of a passing hour' is the expression applied by Thucydides to his own *History* (I. 22).

to his children's children, says Solomon;[1] *children's children* does not mean grandchildren, but remote posterity.[2] Try telling Solomon that 'no one wants capital goods for their own sake ... they are simply the road to future consumption', and so forth.

The second reason is that saving creates wealth by double counting. The same money works twice, once for the borrower and once for the lender. Borrower and lender may be the same person, as in the owner-occupation of unmortgaged property; but the argument is the same whether they are or not.

The saver enjoys the wealth as well as the income it produces so long as he is free to spend the principal; spending power is preferred to spending itself. But if access to the principal is restricted, the restrictions themselves reduce its value, even where there is no desire to spend. There is an analogy with fractional banking. As long as all customers are free to draw their money out of the bank, only a small proportion of the maximum is drawn and the remainder is preferred to cash— a situation which can be stable indefinitely.[3] But if the bank's credit-worthiness comes under suspicion, the customers' fear that they no longer have the power to encash on demand makes them prefer cash to credit. The run on the bank is caused by their fear of no longer being able to do something they do not wish to do if they are sure they can.

The categories of saving and wealth

Saving may be planned or unplanned; short-term or long-term; temporary or permanent; personal or impersonal. The categories overlap.

Short-term saving (for a holiday) and long-term saving (for a pension) are both temporary. Permanent saving may be planned (*a good man leaveth an inheritance to his children's children*); but it is unplanned if it is simply the excess of income over expenditure and becomes permanent merely by virtue of never

---

[1] Proverbs XIII 22.

[2] 'Visiting the iniquity of the fathers upon the children, and upon the children's children, unto the third and to the fourth generation.' (Exodus XXXIV 7.) 'But the mercy of the Lord is from everlasting to everlasting upon them that fear him, and his righteousness unto children's children.' (Psalm CIII 17.)

[3] This argument is not invalidated by the criticism in some quarters that fractional banking causes inflation. It is sufficient for the analogy that wealth is destroyed by a threat to the power to encash, even though this power is little exercised unless it is threatened.

being realised. Planned saving, whether permanent or temporary, is either externally committed (life assurance premiums, mortgage repayments) or at least represents a commitment in the mind of the saver; unplanned saving, whether permanent or temporary, is simply a residue.

Saving is personal if it is directly held by an individual and impersonal if it is held through a chain of intermediaries. A share in a company is personal if it is directly held and impersonal if it is held through a series of financial houses; what is personal here is the direct claim on a financial house. The National Debt is an extreme example of impersonal negative savings.

Similarly for wealth, which is the sum of past saving. Wealth is liquid in an external or banking sense if it can easily be turned into cash and illiquid if it cannot; liquidity in this sense is increased if there is a market. Committed savings are likewise less valuable to the lender and more valuable to the borrower than savings which the lender is free to realise at any time. In general, just as the commercial valuation of encumbered assets (such as a house with an outstanding mortgage) is reduced to take account of the encumbrance, so the valuation of assets for purposes of economic analysis and fiscal policy should be reduced to allow for objective illiquidity and government restrictions and controls.[1]

But there is also an internal or subjective sense in which wealth is illiquid if it is invested in assets which for sentimental or other personal reasons the owner would be reluctant to sell even at a high price and on a ready market. Subjective or internal illiquidity thus implies that the wealth is invested in *personal* assets, which may be defined as those having a significantly higher value for the owner and his family than for strangers. However perfect the market for an asset in dealings between outsiders, the insiders value it well above the market price. A simple example is anything with a high sentimental value (an old photograph perhaps), and no market value at all. Personal assets acquire their *surplus value* through some form of association with the owner or his family or friends—family history, familiarity, work. Important personal assets for which there is a market include firms, farms, homes, furniture, jewellery, works of art.

---

[1] *House of Commons Select Committee on a Wealth Tax, Session 1974-75,* Vol. III, p. 874, para. 22 (Memorandum by the Unquoted Companies' Group).

Although both objective and subjective illiquidity impede the sale of an asset at its full market value, their effects on the value of the asset to the owner are opposite. Subjective illiquidity increases the value to the owner above the market value, sometimes by a large multiple. If the personal asset has no commercial value, the multiple is infinitely large. The asset is illiquid because there is no one who will buy it for what it is worth to the owner. Objective illiquidity, by contrast, depresses short-term market value below the value to the owner—for example, because the market is narrow. Both forms of illiquidity necessarily imply that the value to the owner substantially exceeds the value to anyone else. But subjective illiquidity achieves this result by increasing the value to the owner whereas objective illiquidity achieves it by reducing the value to others. Thus if the illiquidity is subjective, the owner says he will not sell; if it is objective, he says he cannot.

Subjective illiquidity does not imply that the owner will not part with the asset at any price, though this may be true; it simply implies that the value to the owner is higher, perhaps much higher, than the value to anyone else. Where subjective illiquidity is absolute and no offer will tempt the owner, we move out of economics into morality and honour. What is absolute in the sense of this logical extreme will vary from one individual to another. Will any price tempt you to sell your father's Victoria Cross? The saying that every man has his price is not strictly correct but applies to the large area of conduct in which even serious misgivings are not beyond the reach of money.

Anomalies and inconsistencies
These distinctions have been used in policy and in political discussion only fitfully and patchily, when they became too obvious to ignore. Anomalies and inconsistencies abound.

The Royal Commission on the Distribution of Income and Wealth, for example, excluded pension rights from their estimates of wealth ownership on the ground that they are not transferable.[1] Protected agricultural tenancies are similarly excluded; nor are they taxed when they are bequeathed. But life tenancies in trusts, which are no more transferable than agricultural tenancies or pension rights, are included and taxed in full.

[1] *Report No. 1*, Cmnd. 6171, HMSO, Ch. 2, para. 44-46.

Under both estate duty and capital transfer tax, industrial assets have received substantial reliefs in recognition of their external or objective illiquidity. But similar reliefs have also been granted to agricultural assets, whose illiquidity is rather of the internal or subjective kind.

All forms of capital taxation may produce injustices and absurdities when applied to assets which are personal (or subjectively illiquid). The theory of capital taxation, such as it is, is that an increase in capital values implies that the taxpayer is better off and can thus bear a heavier tax burden. But if the asset is personal, this may not be so. If the value to the owner is £500 and the market value is £100, a doubling of the market value to £200 as a result of a halving of the current yield does not make the owner any better off.[1] The tranche of value from £100 to £200 is already pre-empted by the personal character of the asset and cannot be had twice. If capital gains tax is levied on the increase from £100 to £200 or capital transfer tax is levied on £200 instead of £100, the owner is worse, not better, off in consequence of what might at first sight appear to be an improvement in his circumstances. This result is sometimes called 'the farmers' paradox' because farms are a characteristically personal asset. But the paradox is not confined to farmers: any nominal gain may be transmuted into a real loss by the counter-Midas effect of capital taxation if the assets subject to the tax charge are personal.

The farmers' paradox shows the conventional theory of capital taxation to be mistaken not merely in degree but in direction, since the farmer is worse off after tax as the result of what might at first sight appear an improvement in his circumstances; the apparent improvement in his circumstances has no money value for him even gross of tax, given his present situation. It is a main purpose of this *Paper* to argue against the destruction of wealth and welfare by a tax system ignorant of the subtleties of the ownership dimension.

---

[1] The only qualification is that a serious deterioration in the owner's fortunes, whether gradual or catastrophic, may reduce the asset's value to him in the economic sense that its sale is the price of survival. He then gains from being able to sell at £200 rather than £100. But this argument does not apply to a case in which the owner and his heirs never intend to sell the asset and never do so; the power to realise the higher price is then neither enjoyed in advance nor exercised.

Policy towards surplus value

The surplus value created by the personal ownership of personal assets is destroyed if they pass to strangers. Sometimes this destruction happens in the course of nature: if the last of a long line dies without heirs, no one will value the family home as an heir would have done. But more often nowadays this surplus value is destroyed by tax policies intended to have just this effect.

Not only is surplus value destroyed by the transfer of personal assets away from those who value them personally; market value is diminished by their transfer to institutions. Personal assets are unsuited to impersonal saving of which the extreme form is ownership by the state or its agencies. If people's clothes, furniture or homes are owned by government (as in the British system of council housing), the result is wasteful not only because running expenses are increased but also because *no one enjoys the benefit of ownership*; ownership is so widely dispersed among the citizens in their capacity as voters that the benefit is diluted to nothing. Similarly, if works of art are taxed away from their owners and left to gather dust in the basements of museums, there is a double loss: not only does nobody see them, but *nobody owns them either*. The argument that social ownership destroys the surplus value obtainable from ownership proper applies generally to nationalised and 'social' property such as schools, hospitals, railways, power stations and the like.

In a fiscally neutral environment, market forces determine whether or not the potential surplus value that may be realised through the personal holding of personal assets is outweighed in particular cases by other considerations. Some people prefer to be tenants rather than owner-occupiers, for example, and others regard rented property as a good outlet for small savings. A property-owning company can satisfy the requirements of both sides. But the fiscal environment may be far from neutral, and the surplus value obtainable through the personal ownership of personal assets may be destroyed by a biased tax system and other forms of government interference. The worst interferences with the ownership of land during the present century have been planning controls, rent controls, 'leasehold reform',[1]

---

[1] Compulsory transfer of freeholders' rights to leaseholders under the Leasehold Reform Act 1967. Compensation was by definition inadequate (otherwise compulsion would not have been necessary); and it was generally far below any market assessment of freeholders' interests.

compulsory purchase and taxation. Although planning controls constitute an uncompensated removal of owners' rights,[1] they fall outside our present subject since they concern a change of use or development of land rather than its ownership as such. And rent controls are a similar burden on large landowners and small. But the weight of taxation, and its distortion of a neutral pattern, increase drastically as the size of the estate rises.

Other implications for policy

The ownership dimension of welfare and policy should not be neglected, as it has been hitherto. The argument that ownership is diluted or dissipated and ultimately destroyed by dispersal, although it varies from asset to asset as well as from citizen to citizen, is of general application and goes beyond the ambit of the present *Paper*. It is an argument, for example, against nationalised and municipal ownership additional to the arguments about efficiency on current account.

Personal ownership also has a political dimension: it does something to offset the concentration of political power in the hands of the state and its creatures and agencies. It is sometimes suggested that homes and farms, two of the principal 'personal' assets, may have a particular importance for this purpose.[2]

Major economic policies are at present generally assessed by the government and others for their effects on income but not for their effects on wealth. The two ought to be considered together. For wealth, as for income, it is possible—at least in theory—for government intervention to do good instead of harm. If for environmental reasons government compels someone to cease trading in a cathedral close, it is theoretically possible that his losses on current account will be smaller than his neighbours' gains—and similarly for his losses on capital account. Alternatively, the balances on current and capital account may have different signs, one positive and the other negative. Whether the best policy is benign neglect or whether there is a good case for intervention in a particular situation, the ownership dimension forms an integral part of the reckoning.

---

[1] Compensation was envisaged under the Town and Country Planning Act 1947; but it has not in practice been provided.

[2] A. Whitney Griswold, *Farming and Democracy*, Yale University Press, 1948, especially the Preface.

The creation of wealth through ownership should be treated as a major policy consideration, in the sense that it should not be inhibited by taxation without strong reasons. It is not easy to see what these strong reasons might be, since the effect of taxation on the creation of wealth through ownership is wholly destructive and no corresponding gain or gainer can be identified.[1] Rent controls are a different matter. However disagreeable the partial expropriation of one citizen in the interest of another, there is at least a gainer from every such arrangement as well as a number of losers. The political appeal of rent controls is among the subjects discussed in the next section.

---

[1] The creation of wealth through ownership is related to the idea propagated by a number of American libertarian economists—Coase, Demsetz, Cheung and others—of solving or mitigating economic problems through the extension of property rights. (See under 'The creation of property rights as a solution to economic problems', Section III, pp. 48-49.)

II. IS LAND UNIQUE?

Doctrines of differentiation

Does the welfare theory of ownership developed in Section I apply to land as much as to other assets, or has land particular characteristics requiring the theory to be modified?

Within the Western tradition of thought, the idea that land is different from other material assets may be traced back at least to the Hebrew jubile or jubilee.[1] Land was to be restored to its original owners every 49th (or 50th) year.

> 'In the year of this jubile ye shall return every man unto his possession. . . . The land shall not be sold for ever: for the land is mine.'[2]
>
> 'In the year of the jubile the field shall return unto him of whom it was bought, even to him to whom the possession of the land did belong.'[3]

No other material assets were subject to these laws; and the language assimilated the release of land to the release of bondmen.[4] Alienation of land outside the family was originally contrary to the usage of Judaic common law.[5] Freehold alienation was forbidden, and the original owner and his family retained rights as lessor.

Adam Smith and the Physiocrats

An equally sharp differentiation between land and other assets has been made by economists. The Physiocrats (or, as Adam Smith calls them, the Oeconomists[6]) regarded the labour employed on land as the only productive labour. Turgot speaks of the pre-eminence of the farm labourer who produces over

---

[1] Leviticus XXV 8-17, 23-24, 28-33, 50-54; XXVII 16-24.
[2] Leviticus XXV 13, 23.
[3] Leviticus XXVII 24.
[4] Leviticus XXV 28, 41, 54.
[5] 'The Lord forbid it me, that I should give the inheritance of my fathers unto thee.' (I Kings XXI 3.)
[6] *The Wealth of Nations*, Book IV, Ch. IX.

the artisan who manufactures; it is the labourer who makes the earth produce the wages of all the artisans.[1] Quesnay in his *Tableau Économique* contrasts 'productive expenditure relative to agriculture etc.' with 'barren expenditure (*dépenses stériles*) relative to industry etc': 'Productive expenditure is employed in agriculture, meadows, pasture, forests, mines, fishing etc.' Smith, while criticising

> 'The French philosophers, who have proposed the system which represents agriculture as the sole source of the revenue and wealth of every country [for] representing the class of artificers, manufacturers and merchants, as altogether barren and unproductive',

nevertheless considers that

> 'the labour of farmers and country labourers is certainly more productive than that of merchants, artificers and manufacturers'.[2]

Smith indeed makes just as sharp a distinction as the Physiocrats between 'productive and unproductive labour', including in the latter category the labour not only of 'menial servants' but also of 'the sovereign, with all the officers both of justice and war who serve under him' and of 'churchmen, lawyers, physicians, men of letters of all kinds; players, buffoons, musicians, opera-singers, opera-dancers, etc.' The difference between the Physiocrats and Smith is merely that the distinction between productive and unproductive is drawn in a different place.

Smith distinguishes three factors of production—land, labour and capital (or stock); and 'those who live by rent', 'those who live by wages' and 'those who live by profit' are 'the three great, original and constituent orders of every civilised society, from whose revenue that of every other order is ultimately derived'. The interest of the first of these three great orders (rent) 'is strictly and inseparably connected with the general interest of the society' (=of the country); and so is the interest of the second order (wages). But the interest of the third order (merchants and master manufacturers)

> 'has not the same connexion with the general interest of the society as that of the other two. . . . The interest of the dealers

---

[1] M. Turgot, *Réflexions sur la Formation et la Distribution des Richesses* (1788), heading to Section V.
[2] *The Wealth of Nations*, Book IV, Ch. IX.

in any particular branch of trade or manufactures, is always in some respects different from, and even opposite to, that of the public'.[1]

Thus, according to Smith, land differs from other assets both because the labour that works with it is more productive and because those who draw income from it share a common interest with society at large.

Ricardo versus Smith

Ricardo draws the distinction between landlords and manufacturers in the opposite sense: 'the interest of the landlord is always opposed to that of the consumer and manufacturer'. He differs from Smith because 'Adam Smith never makes any distinction between a low value of money, and a high value of corn'.[2] Ricardo was thinking of a situation in which restraints on trade keep the price of corn high; the corn laws were a question of such importance in his day that he treated this situation as general.

A modern economist, the late Professor Fred Hirsch, makes a related but different distinction between land and other assets.[3] Drawing on Harrod's 'unbridgeable gulf' between 'oligarchic wealth and democratic wealth',[4] Hirsch distinguishes similarly between the *material* economy and the *positional* economy. The positional economy is (and the material economy is not) subject to absolute or socially-imposed scarcity or to congestion or crowding through more extensive use. It includes goods, services, work positions and other social relationships. One of his three examples of the positional economy is leadership jobs. But the other two are both forms of land: leisure land and suburban land. The policy prescription he draws from his analysis is 'the principle of restricting certain goods and facilities from private appropriation', through taxation and otherwise.

[1] *The Wealth of Nations*, Book I, Ch. XI, Conclusion.
[2] *On the Principles of Political Economy and Taxation*, Ch. XXIV, Sraffa's edition of Ricardo (Cambridge University Press, 1951), Vol. I, pp. 331-336.
[3] Fred Hirsch, *Social Limits to Growth*, Harvard University Press, 1976, especially Ch. 3, 13/III.
[4] Sir Roy Harrod, 'The Possibility of Economic Satiety—Use of Economic Growth for Improving the Quality of Education and Leisure', in *Problems of United States Economic Development*, Committee for Economic Development, New York, 1958, Vol. I, pp. 207-213.

The Physiocrats and Henry George

Over the last century the two principal threats to personal land ownership in Britain have been taxation and nationalisation. Taxation has, in general, discriminated against land ownership only indirectly: income from the ownership of land is subject to the investment income surcharge, and capital taxes have become an increasingly heavy burden as capital values of land have risen both absolutely and relatively to the values of other assets. Nationalisation, by contrast, is so far not a general threat but a threat to assets held in particular forms. But this contrast is not in the nature of things; Henry George recommended both as elements of the same policy.[1] And George illustrates how analytical distinctions between different classes or factors of production can be just as deadly when their sense is complimentary as when it is critical or hostile. George was no friend of landlords,[2] whereas the Physiocrats were the ultimate theoretical apologists for the landed interest in that they represented 'agriculture as the sole source of the revenue and wealth of every country'. But, when it comes to policy, the Physiocrats and George say much the same: both favour a single tax on rent to replace all other sources of revenue. Indeed, George starts his discussion of the subject[3] with a flattering reference to Quesnay's proposal for a single tax (*impôt unique*).

The analytical questions raised in this section are thus of more than theoretical interest. We now turn to consider how economics can assess and reconcile the doctrinal conflicts noted above between land and other assets and between landowners and other classes.

Assimilation of land to other assets

Section II pointed out how Jevons's marginal utility theory of value complemented Hume's distinction between the pleasure of power and the pleasure of its exercise in providing the theoretical foundations for a welfare theory of ownership. It is also Jevons's value theory that provides the key to the

---

[1] *Progress and Poverty* (1879), Book VI, Ch. II and Book IX, Ch. I.

[2] For example, the title of Book VII, Chapter I is 'Injustice of private property in land' and of Chapter II 'Enslavement of labourers the ultimate result of private property in land'.

[3] *Ibid.*, Book IX, Ch. I.

reconciliation of the various conflicting doctrines differentiating land from other assets.

Jevons generalised the theory of rent propounded by Ricardo and other writers to apply equally to wages and interest, that is, to the rewards of all factors of production:

> 'The parallelism between the theories of rent and wages is seen to be perfect in theory, however different it may appear to be in the details of application. Precisely the same view may be applied, *mutatis mutandis*, to the rent yielded by fixed capital, and to the interest of free capital.'

He thus undid the harm done when 'that able but wrongheaded man, David Ricardo, shunted the car of Economic science on to a wrong line'.[1]

In Jevons's work the distinction between productive and non-productive activities disappears because in a competitive market all activities have the same economic output (utility) and all factors of production receive the same economic reward (the amount determined by the most profitable competing use). Any increase in the reward to capital or labour above this amount is analogous to the landlord's rent. Alfred Marshall used the term 'consumer's rent' in the first edition of the *Principles of Economics* (1890) to denote the analogous concept of the excess over the consumer's spending on a commodity of the amount he would be willing to spend rather than buy less; and 'economic rent' is now a synonym for this 'economic surplus' of producers and consumers alike. This generality is emphasised by the explicit recognition in modern economics that the same agent may combine two or even all three of the rôles of landlord, capitalist and worker. Professor Mark Blaug has written:

> 'For the most part, modern economics has abandoned the notion that there is any need for a special theory of ground rent. In long-run stationary equilibrium, the total product is resolvable into wages and interest as payments to labour and capital—there is no third factor of production.'[2]

But since the real world always falls short of long-run equilibrium, the distinction between land and other assets, driven from the front door, threatens to return through the

---

[1] Jevons, *op. cit.*, Preface to 2nd Edition, pp. lv, lvii.
[2] Mark Blaug, *Economic Theory in Retrospect*, Cambridge University Press, 3rd Edition, 1978, p. 86.

back. 'The distinction between land and other forms of wealth', said Marshall, 'has very little bearing on the detailed transactions of ordinary life.' Nevertheless, 'from the economic and from the ethical point of view land must everywhere and always be classed as a thing by itself'. This is because, 'though the only distinction between land and other factors is that they can be increased in quantity and it cannot; yet this distinction is vital in a broad survey of the causes that govern normal value'.[1]

Land not fixed in quantity
Marshall is right to emphasise the substantial equivalence between land and other assets as alternative investments for the individual investor seeking employment for spare capital. But his argument that land must be classed as a thing by itself because, unlike other assets, it cannot be increased in quantity promotes a difference of *degree*, of which there are many other examples, into a difference of *kind*; and the argument is also wrong in *direction*.

First, the argument that land cannot be increased in quantity is false; in several relevant senses it can. Land is increased in quantity when marshes are drained or polders reclaimed from the sea; about half the present land area of the Netherlands was under water a thousand years ago. Pleasure piers may be constructed in the sea; private citizens built into the water as far back as the time of Horace.[2] Small territories, like Monaco, may build into the sea to ease the pressure on space; they may even build land itself (as in Hong Kong, where the coastline in the Central District is some tens of feet further north than it was at the beginning of the century). More substantially, land is increased in an economic sense by the expenditure of labour and capital on changing its use and improving its quality—clearing scrub, irrigating deserts, constructing skyscrapers—as well as by the use of substitutes such as houseboats for houses. Likewise, land can be moved from place to place, as when topsoil is laid on barren rock.

Secondly, the argument that land is 'a thing by itself' because 'other factors can be increased in quantity and land cannot' is wrong in direction. Although land is *not* fixed in quantity,

[1] Alfred Marshall, *Principles of Economics* (1890), 9th Edition, Vol. II (Notes), pp. 436-437.
[2] Odes II. 18, 20-22.

there are other goods which *are*, such as ancient buildings, works of dead artists and minerals in the ground. But here also, as with land, scarcity and rising prices elicit supplies of substitutes and induce their acceptance.

Land cannot be classified as an asset different from all others because most economically valuable land incorporates other assets in the form of man-made improvements like fertilisers, drains and buildings. Land belongs to the same category of assets as gold and other precious metals and stones, the existing stock of which is many times as large as the annual increment. Their supply can increase; but it does so slowly.

The assimilation of land to other assets extends to the question of development and to 'betterment' and 'worsenment' in general. Rises and falls in land values are no more caused 'by the community' than are rises and falls in other capital values or in wages. The landlord, capitalist and worker are all exposed to many external influences on their fortunes, some beneficial and others harmful, some the work of private men and others of government, some requiring compensation or even prevention and others not. The belief that there should be a separate tax on land and property development has been strengthened by the restriction of development to projects receiving official 'planning permission', a system which may do more harm than good when applied to land and property of no notable beauty or interest.[1] It illustrates the general principle that one form of government interference leads to another.

Elements of difference

Objective balance and subjective imbalance

The persistence of the theory that land is unique is illustrated by the continued support for Henry George's principle of site-value taxation despite its refutation by authors like Seligman.[2] Blaug calls the account of reactions to George given by Cord[3] 'a story of persistent misunderstanding, misrepresentation, and

---

[1] Barry Bracewell-Milnes, 'Market Control over Land-Use Planning', in *Government and the Land*, IEA Readings No. 13, Institute of Economic Affairs, 1974, citing Professor R. H. Coase, 'The Problem of Social Cost', *Journal of Law and Economics*, Chicago, October 1960.

[2] Edwin R. A. Seligman, *Essays in Taxation*, Macmillan, New York, 1895, Ch. III.

[3] Steven B. Cord, *Henry George: Dreamer or Realist?*, University of Pennsylvania Press, Philadelphia, 1965.

downright evasion of issues by the leading members of the economics profession'.[1]

Similar prejudices are reinforced at a popular level by the force of habit and the geography of politics. Most people form an attachment to their workplace, still more to their home, which strengthens with the mere passage of time if the relationship has been a happy one. It requires a strong inducement to persuade them to move, and it is worth something, psychologically as well as financially, to be safe from involuntary removal. These local interests are given disproportionate political influence by a system of territorial political constituencies.

The attachment of the worker or resident to his workplace or home is related to the distinction between personal and impersonal assets and to the attachment of the individual and family to personal assets. These forms of attachment may overlap or even coincide, as in the occupation of private houses by their owners. All this helps explain why some lobbies are more powerful and effective than others.

There are four stages in the argument:

First, there is the contrast between *objective balance* and *subjective imbalance*, which may be illustrated by the example of mortgage finance for private dwellings. For every debt there is a credit, for every mortgagor a mortgagee. The interest paid by the one is income to the other. Apart from differential income taxation, which is not important to the present argument, there is an objective balance between the interests of the parties; whether the interest rate rises or falls, what one side loses the other side gains. But it does not work out like that politically. Politically, a fall in building society interest rates is good news and a rise is bad news, even though mortgagors are fewer and on average richer than building society depositors and have for years been having much the better of the bargain[2]—all circumstances which could generally be expected to reduce their political influence. Part of the explanation seems to be that the personal assets (owner-occupied homes) mean more per pound to the bor-

---

[1] Blaug, *op. cit.*, 3rd Edition, p. 90. Henry George's doctrines have, however, in general been rejected by major economists for more than a century.

[2] T. J. Gough and T. W. Taylor, *The Building Society Price Cartel*, Hobart Paper 83, IEA, 1979.

rowers than the impersonal assets (money savings) mean to the lenders.[1]

Secondly, there are attractions for politicians in forced transfer payments from smaller to larger numbers of voters. The losers may be richer than the gainers; but this is only a minor and indirect influence. The political appeal of rent control and 'leasehold enfranchisement' has been little diminished by the fact that some of the losers have been very poor, some of the gainers (especially from leasehold enfranchisement) very rich, and some of the gainers much richer than the losers whose property rights they were acquiring for nothing. The political importance of the individual's attachment to a familiar place, which is a cousin of the individual's attachment to personal assets, extends beyond homes to the manufacture of property rights in jobs and agricultural tenancies, which may even be inherited and yet enjoy the further advantage of not being recognised as assets for purposes of capital transfer tax. All these property rights are manufactured[2] by an uncompensated transfer from other private citizens, whose rights are correspondingly diminished.

Thirdly, there are attractions for politicians in policies the benefits of which accrue in the short term and the disadvantages only in the long. This helps to explain the appeal, for example, of Keynesian and neo-Keynesian macro-economic 'management' until eventually its cost became intolerable. Rent controls and other forms of the manufacture of property rights fall into this category. The expropriation of one section of the community in the interest of another may have short-term attractions, for the reasons already given; but the poison is more insidious because the long-term damage is concealed or invisible. Rent control may be 'more effective than bombing' in the destruction of cities.[3] But this result cannot simply

---

[1] This is an example of what the jargon of welfare economics calls an *interpersonal comparison of utility*. The validity of interpersonal comparisons has been denied in traditional welfare economics, which is restricted to income and consumption. But, even if the denial is correct, it need not follow that no valid interpersonal comparisons can be made in the ownership dimension of welfare.

[2] The verb 'manufacture' is used in what the Oxford English Dictionary calls the 'disparaging sense' of 'fabricate' or 'invent fictitiously' and is thus distinguished from the *creation* of property rights by an extension of the variety of contractual arrangements or of the opportunities for their use.

[3] Stuart Butler, *More Effective than Bombing: Government Intervention in the Housing Market*, Adam Smith Institute, London, 1979 (also below, p. 100, note 1, and p. 101, note 1).

be seen; it must be demonstrated by argument. The housing shortage caused by rent controls may indeed be regarded as a reason for their continuation at the superficial level of argument that confines itself to immediate effects. 'Between a bad and a good economist', wrote Bastiat, 'the only difference is that the one is satisfied with *visible* effects whereas the other takes account not only of the effects he can *see* but also of those he should *foresee*.'[1]

Fourthly, the election of British politicians in geographical constituencies, in contrast to other systems found elsewhere or conceivable in theory, makes land especially vulnerable to government interference with the market process. Lobbies promote particular interests whereas the market promotes the general interest. The present electoral system manufactures and strengthens lobbies for the promotion of particular interests in land against the general interest served by the market because the beneficiaries of particular interests are geographically concentrated.

We now turn to consider how government interference in the market process affects the characteristics of land discussed so far.

Consequences of government action

The characteristics of land discussed above make it more vulnerable than most other assets to government interference and a more attractive target than most other assets for the activities of lobbies and pressure groups.

Land is not only a *personal asset,* in the sense of being worth more to the owner than to strangers; it is also *proprietary wealth,* in the sense of being worth more to society when it is owned by private persons than when it is owned by the state or its agencies.[2] Thus, there is one loss when land is expropriated and a second if it is transferred to the state or its agencies

---

[1] Frédéric Bastiat, *Ce qu'on voit et ce qu'on ne voit pas* (1850), Foreword. Published with *Premières Notions d'Economie Politique Sociale ou Industrielle* by Garnier Frères/Guillaumin, Paris (6th Edition, 1884).

[2] Whether *most* assets are proprietary wealth in this sense is beyond the scope of this study. But not *all* are. The opposite concept of *communal wealth* (assets worth more to society when owned by the state than by private persons) is exemplified by unlawful weapons handed in under amnesty at the end of a civil disorder. *Surplus value* is generally obtainable not only from a shift from communal to proprietary assets but also from a shift from impersonal to personal assets. It is a distinction of degree rather than kind whether this surplus value is extracted only from heritage assets (Sections V and VI below) or from other assets as well.

rather than to private persons. The fact that land is both a personal asset and proprietary wealth makes it a particularly unsuitable subject for government interference since the interference destroys wealth by transferring property rights from individuals, in whose hands they have value, to government, in whose hands they have none. By contrast, no value is destroyed by contractual re-arrangement of property rights (such as a transfer of rights from landlord to tenant for a consideration). On the contrary, contractual re-arrangements *create* value for both parties since they otherwise would not be undertaken.[1]

The same characteristics of land that make it a personal asset to the owner also make its use and treatment interesting to others—tenants, neighbours and the local community in general. Both the owner and the others acquire a deeper interest and a closer attachment as time goes by. For the owner, this process can go on for centuries; a house owned by a family for 500 years may for that reason alone be worth more to the family than one owned for 400 years. There is no other major[2] category of asset that possesses in similar degree the quality of increasing in value through *continuity* of ownership. Perhaps the nearest parallel is the immaterial possession of a family name. This concept of the creation of value through continuity of possession appeals more to some than to others; it did not, for example, appeal to Juvenal.[3] But, as Jevons pointed out,[4] the judge of value is the individual. If some people

---

[1] There is an overlap between the concept of *proprietary wealth* and that of a *proprietary company*. The latter is defined by the Department of Trade as a small private company which does not exceed two of the following three criteria: (i) turnover: £1,300,000; (ii) balance sheet total: £650,000; (iii) average number of employees: 50 (*Company Accounting and Disclosure—A Consultative Document*, Cmnd. 7654, HMSO, September 1979, p. 1; following similar criteria in the EEC Fourth Directive on Company Law, adopted on 25 July 1978—*Official Journal of the European Communities*, 14 August 1978). But these arbitrary ceilings are at variance with the economic reality that the proprietary quality of wealth need not diminish with the size of the holding. The overlap between the concepts of proprietary company and proprietary wealth helps to explain why there has for at least two generations been a 'Macmillan gap' (shortage of finance for the small but expanding private company) and why entrepreneurs have 'rooted objections to parting with equity almost even in the most desperate circumstances'. (*Institute of Directors Venture Fund*, Spring 1981, pp. 1-2.)

[2] The argument can hold good also for other personal assets such as family portraits and jewellery. But these assets are generally less valuable than houses and land, and the argument often applies to them less strongly.

[3] *Stemmata quid faciunt?* (What is the use of ancient names?) Juvenal VIII 1.

[4] Above, page 17.

value continuity, then it has value even if it is disparaged by others. The task of public policy is to enlarge prosperity, including values that cannot in the ordinary course of events be given a precise monetary magnitude.[1]

The long time-scale that enables the creation of wealth through the ownership of land to realise its full potential is also a characteristic of forestry[2] and of agricultural activities such as the cultivation of fruit-bearing orchards and the breeding of livestock. Most industrial processes have a shorter time span —except for the provision of infrastructural plant like roads, railways, harbours and airports. Governments, by contrast, generally have a time horizon effectively limited to the next election—five years or less. This conflict of interest between government and public makes it rational for government to to behave like Bastiat's bad economist and adopt policies the apparent benefits of which accrue in the short term and the real disadvantages in the long. Rent control is an example.

Thus the *first* consequence of government interference in the land market and the system of land ownership is the destruction of wealth through the transfer of property rights into the hands of the state where they become *sterile*.[3]

The *second* consequence illustrates the general law that one form of government interference leads to another (and, contrariwise, that a reduction of government interference enlarges the scope for a further reduction). Here again rent control supplies an example, since support for council housing (housing provided by government) has been largely due to the housing shortage caused by rent control. But land is affected not only by government policies towards land but also by government economic policies in general, not least by government's success or failure in controlling inflation. Land resembles gold in that each year's production is only a small part of the existing stock. This makes both land and gold good shelters from inflation and much more attractive than the government's own currency even when the currency is lent at substantial rates of

---

[1] This point was clearly recognised by Adam Smith in his discussion of what are nowadays called 'compliance costs' (the costs incurred by the taxpayer in complying with tax law): 'though vexation is not, strictly speaking, expence, it is certainly equivalent to the expence at which every man would be willing to redeem himself from it'. (*The Wealth of Nations*, Book V, Ch. II, Part II.)

[2] Forestry is discussed in Robert Miller, *State Forestry for the Axe*, Hobart Paper 91, IEA, 1981.

[3] Here at least is an apt use for the adjective of Quesnay (above, page 33).

interest. British governments of both parties have resented this competition from superior wares and have reacted with hostility towards both land and gold. Gold was for 40 years subject to restrictions on its ownership, import or export by persons resident in the UK for purposes of exchange control, until controls were abolished in the autumn of 1979.[1] Land has in recent years been subject to increasingly discriminatory taxation in the form of a 'betterment levy' or 'development land tax', even though the rise in land prices in the early 1970s to which these taxes owed much of their political support was caused by the monetary incontinence of government.

One form of government interference intensifies another. Inflation makes land an attractive investment and drives up land prices—which in turn creates the 'farmers' paradox' of a reduction in prosperity through the increase in capital taxation when land prices rise. It also creates the problem of the 'farming ladder'—the difficulty of becoming a farmer owing to the high price of farms[2]—which in turn leads to calls for more intervention by government to resolve another problem of its own creation.

The high prices of farms, and the consequent low yields on the investment they represent, are the work of government. For the cultivator as an individual, said Marshall,

> 'the question whether to cultivate a large piece of land lightly or a smaller piece intensively, is to be decided by business calculations of just the same character as those that govern other applications of his capital and energy'.[3]

But the same argument can be applied to a choice between investment in land and investment in interest-bearing stock; if the one yields much more than the other, the cause must be sought in such differences as relative risk and relative growth prospects.[4] The cause of high land prices is the same debase-

---

[1] Robert Miller and John B. Wood, *Exchange Control for Ever?*, Research Monograph 33, IEA, 1979.

[2] Below, page 71.

[3] *Principles of Economics, op. cit.*, p. 437.

[4] Ricardian rent due to differential fertility is assumed to be capitalised into the price like man-made improvements in the land. Differences in net-of-tax yields due to differences in tax treatment of the returns from different ways of investing the same money are discussed in the Meade Report, *The Structure and Reform of Direct Taxation*, Allen and Unwin for the Institute for Fiscal Studies, 1978, especially Part Two, Ch. 4.

ment of the currency that has made the 'reverse yield gap' the norm since August 1959.[1]

The argument in this *Paper* applies to land in general. Forestry is distinguished by its long time-cycle and by the possibility of using commercial forests for leisure activities. But these are only differences of degree from other forms of agriculture. Uncultivated land raises questions of its own which are discussed in the next section.[2] The classical economists were little concerned with urban rents,[3] and much of the analysis in this *Paper* holds good for urban land as well as for rural. The particularities of urban land ownership form the subject of Section VI.

---

[1] On 27 August 1959, the yield on ordinary shares fell for the first time below the yield on government securities, thereby reversing the historical yield gap that had been maintained by the belief that government securities were less risky. This influence was outweighed for the first time in August 1959 by the fact that ordinary shares, unlike government securities, provided some measure of protection from inflation.

[2] Below, page 63.

[3] 'These writers considered only rural or agricultural rent,' says Buchanan of the Physiocrats, Smith and Ricardo. 'Urban rents were small and of little importance at the time and had no relation to the corn-laws.' (D. H. Buchanan, 'The Historical Approach to Rent and Price Theory', *Economica*, June 1929, p. 139.)

III. SYSTEMS OF LAND OWNERSHIP

The concept of capital-efficiency (maximising the value of ownership)

Economic efficiency is the ratio of output to input. But, just as welfare has traditionally been assessed in terms of income or consumption to the exclusion of ownership (Section I), so economic efficiency has been treated as the ratio of income to expenses per unit of time. This concept may be adequate for the economic efficiency of a machine, whose capital value is determined by estimates of the pattern of its future cash flow. But institutional arrangements, including the tax system, may also be economically efficient or inefficient; and this efficiency ought to allow for capital values which are not merely the capitalisation of income flows.

Economic efficiency in the conventional sense is *income-efficiency*, the maximisation of income relatively to input or when input is given. But there is also the corresponding concept of *capital-efficiency*, the maximisation of wealth. In many contexts the two concepts coincide and no distinction between them is necessary. But it is also possible for higher income-efficiency to be associated with lower capital-efficiency and lower economic efficiency in the aggregate.

Criteria of an economic and fiscal system which maximises the value of ownership

Wealth may itself be more or less income-efficient in the sense of having a higher or lower yield. At any given yield, the four main criteria of a capital-efficient economic and fiscal system are:

(i) treatment of personal assets;
(ii) treatment of proprietary assets;
(iii) marketability;
(iv) ownership intensity.

Personal assets are worth more to their owner than to strangers. Proprietary assets are worth more to society when owned by private persons than by the state.[1] Marketability is liquidity at least over the longer term. Ownership *intensity* is the opposite of ownership dilution or dispersal. These four concepts are discussed in Sections I and II.

Ownership *dilution* must be distinguished from its *division* or *fragmentation*. Ownership need not be diluted at all if a big public company is owned by a large number of small shareholders instead of a small number of large ones, although there is dilution if the company is acquired by the state or one of its agencies. There may also be dilution if a family company is taken over by its employees.

The advantage of ownership intensity may be illustrated by the financing of large public works, such as a harbour, from private saving rather than taxation. Even if income-efficiency and capital requirements are identical for the two financing methods, financing through taxation makes the taxpayers poorer without creating any corresponding assets. Financing through public (in other words, *private*) subscription, by contrast, impoverishes no one since all subscriptions are voluntary. Although the public parts with the same amount of money as when the harbour is financed by taxation, private individuals become richer, not poorer, because they have acquired a proportion of the undiluted ownership of the harbour through their purchase of the shares. This capital value is annihilated by dilution if the harbour is 'owned' by the community at large.[2]

Ownership is at its most intense when assets are owned by single individuals. Maximum intensity is not a prerequisite for the maximisation of wealth; some married couples, for example, prefer to have joint bank accounts or to put the matrimonial home in their joint names. But as dilution proceeds, there is an increasing probability that the welfare available from capital ownership by individuals is diminished and ultimately annihilated. The only exception to this principle is a limited range of public goods such as weapons of war.

[1] The capital-efficiency of personal and proprietary assets illustrates the territorial instinct of many of the higher forms of animal life, including man. (Robert Ardrey, *The Territorial Imperative*, Atheneum, New York, 1966, especially Ch. 3.)

[2] Professor Armen Alchian makes a similar point in contrasting a private theatre with a theatre owned by the municipality. The main difference is that, since the citizen cannot realise his share in the municipal theatre, its value is effectively destroyed. (*Economic Forces at Work*, Liberty Press, Indianapolis, 1977, Part II, Ch. 5, p. 137.)

The creation of property rights as a solution to economic problems
The creation of wealth through ownership is closely related to the concept of solving or mitigating economic problems through the extension of property rights. This applies especially to problems of 'externalities' (uncompensated effects of economic activity on outsiders).

Modern theory on the subject originates in an article by Professor R. H. Coase.[1] Prior to Coase, the conventional doctrine had followed the lines set out by Pigou.[2] Pigou and his followers argued that divergences between private and social costs (for example, where economic activity polluted air or water) required government to impose a tax on the activity to raise private cost to the level of social cost. The argument was extended from taxes to subsidies (for example, where the rest of society had an interest in the training an entrepreneur gave his workforce).

Coase showed that the essential problem was not external effects in themselves but barriers to trade in the form of attenuation of property rights and high transaction costs, including costs of bargaining. If transaction costs were low, the best solution would be reached by bargaining between producer and consumer. Moreover, the best solution was unaffected by whether the relevant property rights were vested in the producer or the consumer. If, for example, the noise from an industrial activity is causing a nuisance to neighbours, the best solution is the same whether the industrialist has the right to make the noise or the neighbours have the right to make him stop; in either case it involves waiving one or other of these rights, wholly or in part, for a cash consideration, and the only difference is the direction in which the money passes.[3]

Property is itself a collection or bundle of rights to various forms of use.[4] The rights can be *partitioned* without being divided or diluted;[5] one person may have the right to walk across a field, another to grow crops there. Rights such as these may

[1] 'The Problem of Social Cost', *Journal of Law and Economics*, Chicago, October 1960.

[2] A. C. Pigou, *The Economics of Welfare*, Macmillan, 1st Edition, 1920, Ch. 6.

[3] A critique of Pigou, Coase and other authors on this subject is given by Steven N. S. Cheung in *The Myth of Social Cost*, Hobart Paper 82, IEA, 1978.

[4] For example, C. B. Macpherson, 'The Meaning of Property', in C. B. Macpherson (ed.), *Property: Mainstream and Critical Positions*, Basil Blackwell, Oxford, 1978.

[5] Armen Alchian, *op. cit.*, p. 132.

be carved out of superior rights which formerly included them —a flexible system for suiting the requirements of both parties. A hotel, for example, can let out its restaurant to a contractor; a department store can rent floor space to individual traders; a barber's shop can even hire out individual chairs to barbers who may be its former employees.

Where there are conflicting interests in the same piece of land, the conflicts can be lessened or even resolved if there is a superior landlord with an interest in the preservation and improvement of the property as a whole. As Professor F. A. Hayek says:

> 'Some of the aims of planning could be achieved by a division of the contents of the property rights in such a way that certain decisions would rest with the holder of the superior right . . . Estate development in which the developer retains some permanent control over the use of the individual plots offers at least one alternative to the exercise of such control by political authority.'[1]

MacCallum puts forward the related idea of a *proprietary community* (hotels, shopping centres, industrial estates, mobile home parks) in which the interests of the various users are reconciled by their common relationship with the proprietor.[2]

We now consider the various forms of land tenure in the light of the opportunity they offer for the creation of wealth and the resolution of problems through the extension and articulation of property rights. Property rights may be articulated or extended through the development of law or custom which creates private rights out of rights formerly annihilated by dilution or anarchy. Professor Harold Demsetz, for example, drawing on work by Eleanor Leacock, cites the development of property rights among the Indians of Labrador in response to the development of the fur trade.[3]

---

[1] F. A. Hayek, *The Constitution of Liberty*, Routledge & Kegan Paul, 1960, p. 352; also below, page 103 note 3.

[2] Spencer Heath MacCallum, *The Art of Community*, Institute for Humane Studies Inc., Menlo Park, California, 1970. Professor D. R. Denman uses the related concept of the *proprietary land unit* to mean 'an area of land managed as a single entity and co-extensive in its physical dimensions with vested rights of user, disposition and alienation'. (*Land Use and the Constitution of Property*, Cambridge University Press, 1969, p. 25.)

[3] Harold Demsetz, 'Toward a Theory of Property Rights', *American Economic Review*, May 1967, p. 347. Eleanor Leacock in *American Anthropologist* (American

[*Contd. on p. 50*]

Market and state: a schematic classification of landholding systems

'It would be exceedingly difficult', says Professor D. R. Denman, 'to identify and classify all tenurial systems in existence.'[1] No such attempt is made here. But a brief, modified version of Denman's schematic classification is used to list the principal forms of landholding and assess them by the criteria developed earlier. The landowner generally may, and sometimes must, double the rôles of landlord and tenant by farming the land himself.

The rights of a landholder are absolute if they are not derived from another party's right, and derivative if they are carved out of a superior interest. But this distinction may sometimes be formal rather than substantial. For example, all land in England is held by the Crown, and all other interests, including freehold, are derivative. But in practice there is a sharp distinction between Crown Lands, to which the rights of the Crown are substantial, and other lands, to which the rights of of the Crown are merely nominal. There is little or no distinction between an English freehold interest and a corresponding absolute interest in other countries. Interests like this which are only nominally derivative may be called *autonomous*. Other derivative interests are *reciprocal* (English landlord-and-tenant system), *dependent* (share-cropping, *métayage*), or a *right to use* (licence to occupy). And, just as the position of the freeholder may strengthen relatively to the Crown, so the position of the tenant may strengthen relatively to the private landlord (Britain) or the state (India)—the tenant even enjoying the right to bequeath his tenancy without charge to death taxes.

1. *Private property: single ownership*
   (a) Individuals (nationals or foreigners, as also for the following categories): according to the intensity of use, rural holdings vary from peasant proprietorships through family farms to extensive holdings (*latifundia*).

[*Contd. from p. 49*]

Anthropological Association), Vol. 56, No. 5, Part 2, Memoir No. 78. The article by Demsetz is reprinted in Eirik G. Furubotn and Svetozar Pejovich (eds.), *The Economics of Property Rights*, Ballinger Publishing Company, Cambridge, Mass., 1974, which is the most substantial work on the subject to have appeared so far.

[1] D. R. Denman, *The Place of Property*, Geographical Publications, 1978, Ch. 8, p. 101.

(b) Pluralities: different individuals have rights to different entities on the same land: one person may, for example, have a right to the fruit of certain trees only.

(c) Trustees of a discretionary trust or of a trust with interest in possession.

(d) Institutions: a category covering a wide range, from financial institutions through Oxford and Cambridge colleges to religious houses whose interest in land may resemble a single owner's more than a financial institution's. Financial institutions include insurance companies, pension funds and charities. Their holdings and purchases are generally of prime land and scarcely at all of marginal and hill land.

2. *Private property: communal ownership*

(a) Service co-operative: in West European co-operatives the production units are often owned individually, while the service areas and activities are owned and run co-operatively.

(b) Severalty-cum-common: the ancient form of tenure in which some farming lands are owned individually and others communally, the individual farmers enjoying such rights as pasturage. The communal principle here affects production rather than services.

(c) Periodic re-allocation: individual rights to farming land are re-allocated at intervals.

(d) Partnerships: the identity of the partners' shares is merged as long as they remain partners, although a small family partnership such as that between father and son may be much more like 1(a) than 2(a)-(c), especially if the shares of senior partners are destined to be bequeathed to their juniors. And in some countries partnerships have legal personality, so that they would be better classified under 1 than under 2.

(e) Proportional proprietorship: a form of collective in which the assets are owned by the co-operative which is in turn owned by its members. A limited liability company is an example of a co-operative of this kind in which income is strictly determined by input and the member is free to withdraw his share. A controlled unquoted company may be a vehicle for ownership by an individual. If a member is not free to withdraw his share, the distinction between return on capital and remuneration for labour is obscured or obliterated.

(f) Lost identity: finally, as the culmination of *(a)* to *(e)*, the member may lose his right to a proportional share even during the period of his membership, as in the Israeli kibbutz. All forms of property may be abandoned, even clothes, for instance, being owned communally.

(g) Tribal: communal ownership among people of common lineage, subsidiary rights being available to individuals within the tribe.

3. *State property*

(a) National government: Denman has pointed out that land nationalisation as commonly understood would affect only the superior interest in land. The citizens' requirements for the use of land would have to be met by derivative interests. The derivative interests would themselves be the new commodity, in a new land market dealing with them as a form of private property.[1] But the state can appropriate and retain the subsidiary interests as well, as in the Soviet *sovkhoz*.

(b) Local government: comment as for 3*(a)*.

(c) Derivative interests: state authorities may grant subsidiary interests, not only to their own creatures and agencies, but also to private persons.[2]

---

[1] D. R. Denman, 'Land Nationalisation—A Way Out?', in *Government and the Land, op. cit.*, p. 45. A similar conclusion can be reached by a Marxist route: 'Total land nationalisation is therefore not an aim to be fought for as an aim in itself . . . It must be fought for, but in the knowledge that its primary effect will be not to end the struggle over land-related issues, but to change the conditions of that struggle.' (Doreen Massey and Alejandrina Catalano, *Capital and Land: Landownership by Capital in Great Britain*, Edward Arnold, London, 1978, p. 190.)

[2] In so far as the superior interest of government is substantial, not merely nominal, and the derivative interest is therefore not autonomous, the monopolisation of superior interests in government hands poses problems that are no less serious politically than economically, although only the latter aspect falls within the subject of this *Paper*.

4. *Joint ventures*

   (a) Vertical: lease of land by the state to a private person or *vice versa*.

   (b) Horizontal: partnership between the state and a private person where each has a similar interest in a different piece of land.

5. *State control of private property*

   (a) Restrictions on the acquisition and retention of land:
   primogeniture, until its abolition in England and Wales in 1926;
   compulsory division among legal heirs, the opposite of primogeniture and a main reason why there are so many small farms in continental Europe;
   restrictions on foreigners;
   ceilings on the holding of land by any one individual.

   (b) Transfer of rights by the state from one group of private persons to another: Solon's *seisachtheia*;[1] redistribution of Saxon lands by William I; redistribution of Church lands in the 16th and 17th centuries; control of domestic and agricultural rents; 'leasehold enfranchisement'.

   (c) Government measures for directing land use: zoning; planning controls;[2] government spending.

   (d) Eminent domain: power of compulsory acquisition in what is alleged to be the 'public interest' for what is alleged to be 'just compensation'. (This compensation is by definition below the value of the property to the owner since otherwise compulsion would be unnecessary.)

---

[1] Exoneration of debt through discharge of encumbered land from its encumbrances and re-instatement of persons who had been sold as slaves or had gone into exile for indebtedness.

[2] Zoning and planning controls are negative in the sense that their purpose is to prevent people doing what they wish with their own property. It is arguable that the destruction of wealth caused by this veto may be more than offset by other benefits. But there are cases where this is clearly not so. Restrictions on the use of a listed building, for example, may prevent its being used for an alternative purpose that would generate the funds required for its upkeep but do nothing to prevent its falling down for want of maintenance.

Northfield and CLA classifications

Not all these categories are important in the British context. The Northfield Report[1] gives the following figures for agricultural land ownership in Great Britain in 1978:

	Million hectares	% of total
Public and semi-public bodies and traditional institutions	1·5	8·5
Financial institutions	0·2	1·2
Private individuals, companies and trusts (by subtraction)	16·0	90·3
	17·7	100·0

No figures are available analysing the 90·3 per cent owned by private persons into its component categories. But the Report on the Wyre Forest Survey by the Ministry of Agriculture[2] gives the following breakdown of ownership in the Wyre Forest:

	Hectares	% of total
Individual	4,666	53·5
Partnership and joint ownership	1,740	20·0
Trusts	629	7·2
Private and public company	481	5·5
Central and local government	1,015	11·6
Charities	190	2·2
	8,721	100·0

Additional categories mentioned by the Country Landowners' Association (CLA) in its paper 'Developing Forms of Landownership'[3] include: trustees/beneficiary of a discretionary trust; trustees/life tenant/remainderman of a settlement with an interest in possession; quoted company; controlling/non-controlling shareholder in an unquoted company; overseas landowner; insurance company; pension fund.

[1] *Report of the Committee of Inquiry into the Acquisition and Occupancy of Agricultural Land* (Chairman: Lord Northfield), Cmnd. 7599, HMSO, July 1979, Table 12.
[2] Northfield Report, *op. cit.*, Table 11.
[3] A.556 of 14 July 1977.

There are indications that the ownership of land by financial institutions is of growing importance. In the year ended 30 September 1978, financial institutions enlarged their holdings of agricultural land in England by about 10 per cent of the acreage changing hands.[1]

The income-efficiency of landholding systems

It would be beyond the scope of this *Paper* to assess the efficiency of the various forms of landholding in Britain for generating income year by year. Our purpose is rather to spell out the implications of the arguments in earlier sections and to illustrate some of the conclusions with figures. Income-efficiency is not everything. And even if the fiscal and other legal environment were neutral between the various forms of landholding, there might still be significant differences in income-efficiency because some landowners retained land which they used less productively than alternative owners since it conferred compensating advantages. In practice the environment is not neutral. Differences of this character are compounded by the distortions in the fiscal and economic system, as also perhaps by the lack of an institutional framework enabling the market to function effectively.

There are four principal considerations determining the relative income-efficiency of landholding systems. *First*, where there is no security of tenure, the tenant has no incentive to invest in the land. This has not been a problem in England for hundreds of years. 'In England', wrote Adam Smith, 'the security of the tenant is equal to that of the proprietor.' Indeed, the legal position of the tenant was fortified by custom:

> 'There is, I believe, nowhere in Europe, except in England, any instance of the tenant building upon the land of which he had no lease, and trusting that the honour of his landlord would take no advantage of so important an improvement.'[2]

Current British legislation[3] has had the effect of granting the tenant security for life at a rent below the market-clearing price, and the most recent addition to this legislation[4] has

---

[1] Northfield Report, *op. cit.*, Table 14, p. 65.

[2] *Wealth of Nations*, Book III, Ch. II.

[3] Part III of the Agriculture Act 1947, consolidated by the Agricultural Holdings Act 1948 and amended by the Agriculture Act 1958.

[4] Part II of the Agriculture (Miscellaneous Provisions) Act 1976.

extended the security to as many as three generations including the present occupier. There has consequently been a premium for vacant possession amounting to more than half the price of untenanted land in the late 1940s and about a quarter in September 1978 in England and Wales. The figures are higher for Scotland.[1] The results are comparable with those of domestic rent control. New entrants to farming are kept out, and tenants who could farm the land more profitably are denied the opportunity. A better balance is struck by legislation like the Landlord and Tenant Act of 1954 which provides security of tenure for commercial and industrial tenants subject to the payment of a market rent.

Secondly, communal ownership removes the incentive to economise on scarce resources unless the owners are united by strong common beliefs, as in a monastery or kibbutz. It was the economic superiority of private over communal ownership that provided the motive for the English enclosures in the 18th century and earlier. Similarly, Tripolitania and Cyrenaica (the modern Libya), once the granary of Rome, were reduced to desert when the Roman system of private property rights was replaced by communal rights under the Vandals, Berbers and Arabs.[2] The same argument applies to economy in current use and to the conservation of land for use by future generations. 'An owner of a private right to use land', says Demsetz, 'acts as a broker whose wealth depends on how well he takes into account the competing claims of the present and the future. But with communal rights there is no broker.'[3] Considerations like these may help to explain why co-operative production has made so little headway in British farming and also why private plots in the USSR have made a contribution to agricultural production out of all proportion to their size.[4]

The weakening or removal of the incentive to economise under communal ownership is part of a larger problem of group decision-making. Denman's 'first law of proprietary magnitudes' is that the degree of competence with which the power of decision-making inherent in the property power is

---

[1] Northfield Report, *op. cit.*, p. 96.

[2] John Burton in *The Myth of Social Cost, op. cit.*, p. 87.

[3] H. Demsetz, *op. cit.*, p. 355.

[4] Karl Eugen Wädekin, *The Private Sector in Soviet Agriculture*, University of California Press, 1973, especially Tables 1, 8, 18.

used moves in inverse ratio to the number of joint owners.[1] This law operates most strongly under government ownership, that is, administration by officials in the nominal interest of the whole population.

Thirdly, incompetent decision-making under communal and government ownership is related to the additional cost of administration, not least through the imposition of additional administrative tiers. The reason is the same: it is more costly to manage someone else's property than one's own, because costless self-interest is replaced by external controls. This distinction is found at its weakest (or not at all) in the relationship between an agent or factor and a single landlord, sharing common aims and interests and working closely together for a generation or more. It is at its strongest in management of land by political committees through the agency of officials on short-term assignment. The Ministry of Agriculture Enquiry into Expenses of Landownership[2] gave the following figures for management expenses on let land in 1973-4:

	£ per acre
Individuals and private	0·79
Public and semi-public	1·04
County councils	2·48

Fourthly, there is evidence that income-efficiency increases with farm size, at least to a certain level. Britton and Hill, for example, suggest efficiency gains up to 600 acres in dairy farming and up to 1,500 acres in livestock farming.[3] Such figures are bound to be contentious; and the truth will vary over time and space. But if there is an optimum economic size as well as a minimum size for economic survival, it indicates the cost to the economy of a tax system which makes it increasingly difficult for farms in private hands to attain the economic optimum.

These criteria suggest that the system of tenure which maximises current profit (and thus income-efficiency) is private ownership with security of tenure at a market-clearing rent and

---

[1] D. R. Denman, *The Place of Property, op. cit.*, p. 40.

[2] *Expenses of Agricultural Landownership in England and Wales 1973-1974*, Ministry of Agriculture, 1976.

[3] Professor D. K. Britton and Dr N. B. Hill, Agricultural Economics Society Conference, December 1978.

with owned or tenanted farms large enough to realise the potential economies of scale.[1]

The capital-efficiency of landholding systems

Since one quality of wealth is to constitute the present discounted value of future income, the capital-efficiency of landholding systems is in part the mirror-image of their income-efficiency. Thus the poor husbandry and higher costs resulting from communal and government ownership are necessarily reflected in capital values as well as income returns. But capital-efficiency has another dimension: the pleasure of ownership for its own sake, even at the cost of income forgone. Many hill farmers and others on the margin of self-sufficiency, for example, could earn more in towns; but they stay on the land because the farms are theirs.[2] Another example is the unremitting struggle by the owners of numerous historic houses to keep them going as private homes: 'You don't own this place; it owns you', was the remark made to one such owner by a visitor. It is the pride or pleasure of ownership, often fortified by feelings of obligation towards the family and the local community, that persuades owners to forgo the easier life they would have by selling out and moving elsewhere.

Capital-efficiency may thus conflict with income-efficiency if proprietors prefer to stay on small and marginal farms rather than sell out and leave. But their conduct is not irrational, as is sometimes thought when welfare is assessed only in terms of income. When the ownership dimension of welfare is taken into account, they are better off with their present low incomes than if they moved to higher-paid work elsewhere. That many

---

[1] Security of tenure at a market-clearing rent balances the interests of landlords, sitting tenants and potential tenants and can be attained without government intervention through various forms of contract between the landlord and tenant (or leaseholder). Provisions in the contract can cover the compensation of a leaving tenant for improvements such as new buildings. The view of Adam Smith quoted above (page 55), however, indicates that the discouragement of good husbandry through insecurity of tenure was not a problem in his day. Nor has it been since. On the contrary, artificial security of tenure (security at far below a market-clearing rent as a result of government interference) has been a growing problem in recent years. Artificial security in this sense reduces not only the incomes of landlords and potential tenants but also the capital value of the land subject to control.

[2] Marginal farmers include tenants attracted by the way of life even at the cost of higher income forgone. But the incentive of ownership is additional to the attractions of the way of life.

farmers are willing to make this apparent sacrifice indicates the welfare potential of personal landownership.

Derivative interests may generate welfare through ownership as well as superior interests. In a well-ordered market system the total bundle of rights is divided between the interested parties in accordance with what each is willing to pay for the rights he wishes to have. The superior interest may be little more than nominal. The purpose of building and selling flats on long lease rather than freehold is primarily to have someone more substantial than a residents' association to provide necessary common services; the rôle of the lessor may be confined to providing these services at cost. This is an example of MacCallum's proprietary community in which the rôle of the proprietor is reduced to that of a manager and the owners of the subsidiary rights enjoy the welfare generated by ownership.[1] The lessees have not obtained their position at the expense of the landlord, but as a result of a business deal. Holders of domestic and agricultural controlled tenancies, by contrast, have obtained their manufactured rights of ownership at the expense of the landlord and have thus diminished the value *of a personal asset*. Other tenants, who would be willing to pay more to rent the same home or farm, are also the losers. Under a well-ordered market system, legislation should not at most go beyond providing security of tenure subject to payment of a market rent. Even for this purpose, legislation is unnecessary since the same result may be achieved by bargaining between the parties.[2]

In a well-ordered market system, the division of rights to the same property between two or more citizens can create more wealth than if the law conferred all the rights on one to the exclusion of the other. Just as the same money can work twice, once for the lender and once for the borrower, so the same property can work twice, once for the landlord and once for the tenant. The intervention of government in this process and its manufacture of rights for one party at the expense of the

---

[1] Above, page 49.

[2] Above, page 58, note 1. Security of tenure imposed by legislation instead of being achieved by a contract agreed between the parties necessarily inflicts a loss on the landlord by restricting his freedom of action. If the landlord is free to fix the rent, the loss may be purely psychic (inability to get rid of an unwanted tenant). If the rent is fixed by third parties, the landlord also loses financially from the application of the principle of eminent domain (5d above, page 53.)

other *sterilises* some of the potential wealth inherent in the ownership of property.[1]

Another form of sterilisation is the movement of assets out of personal ownership into trusts. The essential principle of a trust is the separation of beneficial interest from control. The separation diminishes welfare by suspending some of the advantages of individual ownership, though this loss may be more than offset by the advantages of safeguarding the continuity of family ownership against a variety of hazards affecting individual family members. Trusts are a long-standing institution in the English-speaking world. The fact that settlors have been willing to forgo some of the advantages of individual ownership demonstrates that trusts can confer benefits outweighing the loss (and this was so long before the avoidance of taxes on death became a serious consideration). The potentialities of trusts for tax avoidance have been seen in some quarters as a loophole needing to be blocked. But, in so far as they have been used to reduce capital taxes, the problem is rather the wastefulness of a tax system that induces taxpayers to accept an inferior substitute for outright ownership as the price of not paying an even higher tax bill. This wastefulness would continue under the present system of capital taxation even if trusts were proscribed or taxed out of existence. It would merely be hidden.

The four criteria of capital-efficiency

The four criteria explained in Sections I and II for assessing the capital-efficiency of landholding systems are: marketability (or liquidity); the treatment of personal assets; the treatment of proprietary assets; and ownership intensity (the avoidance of loss through the dilution of ownership). Although the criteria apply to assets in general, they are particularly relevant to land and real property. Marketability, ownership intensity and the treatment of proprietary assets are related to the incentive to economise and the competence of decision-taking, which are criteria of income-efficiency. The treatment of personal assets, by contrast, is independent of income-efficiency, since the emotional capital invested in the asset is personal to the individual owner.

---

[1] The *vacant-possession premium* in England and Wales varied between 13 and more than 50 per cent over the period 1945 to 1977. The premium in Scotland in 1977 was 37 per cent. (Northfield Report, *op. cit.*, p. 96.)

Government ownership of land, at national, regional or local level, is inefficient by all four criteria of capital-efficiency. Wealth is destroyed because the assets cannot be realised by their nominal owners, the people, whose interest in the capital value of their nominal property is annihilated by dilution. Such qualities as distinguish land from other forms of wealth make it a proprietary asset like personal and family chattels. And the additional wealth available through the holding of proprietary assets by individuals is destroyed if they are owned by the state. Further wealth is destroyed if land or other personal assets are acquired by the state or transferred to another owner through state action, as when family portraits are acquired in lieu of death taxes.

Co-operatives lose more and more in capital-efficiency as they move further and further away from a purely servicing activity, such as the provision of common marketing. There is a variety of different patterns, some with a strong admixture of state coercion. But one of the main criteria distinguishing the more capital-efficient producer co-operatives from the less is whether the member buys his way in and sells his way out, as in a genuine partnership.

Civic and national pride may act as a substitute for the satisfaction of undiluted ownership, especially in small and cohesive communities. But such sentiments provide few arguments for government ownership, since they may be no less strong if the assets are owned privately. Citizens may take pride in a church of which they are not members, in a historic house in private ownership, in a school or hospital run privately, or in local industry or commerce. At national level, there is relatively less scope for civic pride in government-owned assets simply because nothing will be known about most of them by the population in whose name they are held.

A more effective substitute for personal ownership than civic or national pride is the feeling of belonging to a real community, one whose identity is not merely geographical. The longer and deeper the members' commitment to an institution, the more effective its substitution for personal ownership. Little of the surplus value inherent in proprietary assets may be lost if they are owned by a school or college—still less if they are owned by a monastery. The result depends on the reality and importance of the community in the minds of the members; it can change while outward forms remain the same. The original pioneering spirit in the kibbutzim has not proved easy to maintain as Israel has become more prosperous and secure.

Ownership of land by financial institutions provides marketable assets and need not diminish ownership intensity. But land and other proprietary assets are more than just investments, and this additional dimension is lost if they are owned by financial institutions.

Unquoted companies, though nominally financial institutions, may in reality be little more than a projection of the personality of their proprietor. Controlling interests may be substantially equivalent to individual ownership. Minority interests, by contrast, may be difficult or impossible to sell. The usual principle applies: the illiquidity of minority shareholdings in unquoted companies constitutes no loss of wealth in an economically neutral environment because the situation is by definition accepted voluntarily and without coercion; but wealth is destroyed if minority interests are favoured by legislation relatively to controlling interests.

Personal ownership satisfies the four criteria of capital-efficiency better than any other form of landholding (although, even in an economically neutral environment, capital-efficiency may on occasion be outweighed by other considerations). Controlling interests in companies and family partnerships may be substantially equivalent to individual ownership.

This list exhausts the Northfield and CLA classifications. Charities, the remaining category, are in varying degrees communities or financial institutions or both. The final category in the classification of landholding systems, state control of private property, is by definition an encroachment on property rights imposing additional costs, and it can be justified only if it confers benefits outweighing the disadvantages. The argument that zoning and planning controls do more harm than good goes beyond the ambit of this *Paper*.[1] Why the interests of amenity and conservation may be better safeguarded by private ownership without the intrusion of government control is discussed in the following section.

There are few forms of land use to which these arguments do not apply. The exceptions are a narrow band of uses concerned with government's central functions of defence, policing and the administration of justice. Even here, there is scope for

[1] 'Market Control over Land-Use Planning', in *Government and the Land, op. cit.*, pp. 73-95.

private landownership; firing ranges, for example, may be rented from private landowners rather than bought outright by government. Outside this band, there need be little restriction on the private ownership of land. National parks, for example, which US precedents might suggest as suitable for ownership by government, are privately owned in Denmark and Britain.

> 'Americans find it hard to visualise a national park of hundreds of square miles with virtually no land owned by the national park authorities; but this is what the ten National Parks of England and Wales consist of, and they cover 9 per cent of the country.'[1]

And this system is effective for its own purposes as well as creating the wealth inherent in private landownership: 'These factors have promoted a system which secures the efficient and intense utilisation of a high-quality resource'.[2] The wealth inherent in private ownership may be substantial even when the assets are of little or no commercial value (and even though the restrictions entailed in the establishment of national parks reduce the wealth of the owners). If strangers are willing to pay for the mere title of lord of a manor with which they have no previous connection, it is not surprising that substantial value can be created by the holding of even waste lands in the form of personal assets, especially where there is a long-standing connection with the family.[3]

Land reform

The concepts of income-efficiency and capital-efficiency also help to classify the various kinds of land reform by their essential characteristics.

---

[1] Warren A. Johnson, *Public Parks on Private Land in England and Wales*, Johns Hopkins, 1971, p. xii.

[2] Johnson, *ibid.*, p. 111.

[3] The argument that wealth can be created by the private ownership of proprietary and especially personal assets, even when these assets yield no monetary or other tangible return, is general and not confined to land. It applies, for example, to the acquisition and retention of cherished car numbers. The scope for wealth creation through matching car number plates with car owners is indisputable, even though it may be arguable whether or how far this form of wealth creation should be subject to fiscal or other government constraints. A discussion document which recognised the scope for wealth creation through matching cherished numbers was issued by Mr Norman Fowler, the then Transport Secretary, in March 1981. (*Cherished Numbers*, D9/10, Driving and Vehicle Licensing Centre, Longview Road, Swansea SA6 7JL.)

Elias Tuma has offered an analytical scheme of land reform based on the distinction between individualist and collectivist reforms (his Class I and Class II respectively).[1] Although 'no reform has perfectly fitted in either' of the two classes, Class I is an idealised version of reforms in Western and developing countries and Class II of reforms under communism. Thus the dominant theme of Class I reforms is the promotion of family farms and of Class II reforms the elimination of private tenure.

A more comprehensive classification would distinguish between the *transfer* of property rights between citizens through the agency of state compulsion and the *increase* or *reduction* of private property rights in total through the reduction or increase of collective ownership by the state. The sum of existing private property rights is diminished if they are compulsorily redistributed between citizens. But this process is in principle distinct from an outright extension of public ownership. Moreover, the *transfer* of property rights between citizens may be combined with their *creation* through the replacement of communal by individual ownership, as in the English enclosures, so that capital-efficiency need not be reduced.

Tuma assumes that Class I reforms are intended to 'reduce rural wealth and income inequality' and thus benefit small working farmers at the expense of large landowners. He also assumes that Class II reforms are always in the direction of collectivism. This is unnecessarily restrictive. Transfers between citizens may take not only the traditional form of redistribution from large landowners to peasants and small farmers but also its opposite, as in the English enclosures. Shifts in the boundary between private and state ownership may be not only to the advantage of the state, as in the Soviet Union and other communist countries, but alternatively to the advantage of its citizens, as in the sale of government land to private owners in Taiwan during the years 1948-58.[2] Another example is the the reclamation of the polders by government in the Netherlands and the creation of private derivative interests out of the superior property rights of the state in the reclaimed land.

Land reform in this broad and neutral sense is more a matter of capital- than of income-efficiency. Economic motives (in

[1] Elias H. Tuma, *Twenty-Six Centuries of Agrarian Reform*, University of California Press, 1965, Ch. XIV.

[2] Russell King, *Land Reform: A World Survey*, G. Bell and Sons, London, 1977, p. 211.

the sense of income-efficiency) are outweighed by social and political considerations (including capital-efficiency) in the discussions of the objectives of land reform by Tuma, Russell King and Doreen Warriner. 'Prior to the Soviet reforms, economic objectives were not common features of the reforms under study.'[1] And capital-efficiency is itself a determinant of income-efficiency. 'The weight of the evidence', says Warriner in discussing the economic effects of reform, 'is in favour of granting ownership to individual landholders or rights to individual land use within a group holding ownership.'[2]

---

[1] Tuma, *op. cit.*, p. 180.

[2] Doreen Warriner, *Land Reform in Principle and Practice*, Clarendon Press, Oxford, 1969, p. 434.

IV. EFFECTS ON THIRD PARTIES

Principles governing the public interest in landholding systems
Section III was about the effect of the various landholding systems on the interests of the owners. This section is about their effect on others, both particular groups and the public in general. As in other economic relationships, the effects on outsiders may be adverse or beneficial. Public policy must take account of these 'externalities', not least because of the citizens' interest as voters, and try to promote the joint interests of the owners and outsiders when they are in harmony and strike a balance between them when they are in conflict. In economic theory and policy-making there has been a predisposition to concentrate on adverse externalities.[1] But the external effects are in general just as likely to be beneficial, and it is argued below that the interests of landowners and society are usually in harmony.[2]

Common observation and the analysis in Section III suggest five general principles governing the public interest in systems of agricultural landholding ('general' in the sense of reflecting a consensus, without any implication that everyone does or should agree).

First, in the country resident landlords are generally preferred to absentees.[3] This is partly because tenants and employees can have direct access to the boss and partly because the landowner can make a major contribution to the life of the community. The same principle applies even down to single private homes. Residents are generally more welcome in villages than the owners of holiday homes; Plaid Cymru has included restrictions on second homes in its political programme. The British tax system works against this principle when rural

---

[1] For example, Earl O. Heady and Larry R. Whiting, *Externalities in the Transformation of Agriculture*, Iowa State University Press, 1975.

[2] The particular considerations affecting urban estates are discussed in Section VI.

[3] 'Villagers of Thimbleby in North Yorkshire have learned that they will have a real landlord living in the Georgian manor house instead of the faceless financial institution they had all feared.' (David Hoppit, *Daily Telegraph*, 19 September 1981.)

houses forced onto the market by taxes on capital are sold to townsfolk.

Secondly, whatever they may believe as businessmen or employees, most people in their capacity as residents believe all change is for the worse. The principal consideration that makes planning control acceptable to the public is its conformity to this prejudice. And the prejudice seems to be increasing in intensity. The gestatory period for a new motorway, for example, is about twice as long as a generation ago, mainly because the opposition of residents is more intense. The form of rural landownership which can best satisfy these feelings, namely, family ownership without substantial change of use from generation to generation, is precisely the one most heavily penalised by taxation—specifically by the combined weight of taxes on income and capital.

Thirdly, there is the related distinction between the short and long term. This is the opposite of the distinction between the 'short-term' private interest and the 'long-term' public interest which has been used to justify pouring 'long-term investment funds' amounting to billions of pounds down the drain of the Industrial Reorganisation Corporation, the National Enterprise Board and the nationalised industries. In a purely business context, the short and long term must take care of themselves. Projects which cannot attract private capital should be strangled at birth, and the suggestion that long-term ones should have priority is a prescription for wasting public money. But in a rural community, a financially less attractive long-term plan may indeed be preferable to a financially more attractive alternative, especially for the rest of the community if the cost is borne by the landowner.[1] Here again, it is just this form of private initiative which is inhibited by the tax system.

Fourthly, in so far as political economy can offer a theory of envy, it suggests that the objects of envy are the slightly richer citizens rather than the much richer.[2] The same argument suggests that the opposite of envy, namely, goodwill and the feeling that people ought to be able to live well, is more likely

---

[1] Below, page 73, note 1.

[2] Harry G. Johnson, 'The Macro-Economics of Income Redistribution', in Alan T. Peacock (ed.), *Income Redistribution and Social Policy*, Jonathan Cape, 1954, p. 26; Sir Keith Joseph and Anthony Sumption, *Equality*, John Murray, 1979; Helmut Schoek, *Envy: A Theory of Social Behaviour*, Secker and Warburg, 1969.

to be extended towards the much larger estates than the slightly larger.

Fifthly, these sentiments and interests are not the exclusive concern of neighbours. Charities that oppose whaling attract support from people who are unlikely ever to see a whale; and their support is not due primarily to the cruelty of whaling (which has always been barbarous) but to a now fashionable concern for conservation. The National Trust sends its members each year an appeal for particular properties, named, described and illustrated; the pattern of support indicates that a substantial proportion of the funds comes from people who are unlikely ever to visit the properties in question.[1] Such voluntary actions by large numbers of subscribers show that the public interest in preserving nature, art and history goes far beyond the immediate personal concerns of the citizens.

The 'external' or 'third-party' arguments reinforce the 'internal' arguments in Section III in favour of devolving land-use rights to the individual citizen. Of the four criteria of capital-efficiency (page 60), three are the same for the public as for the proprietor: treatment of personal assets, treatment of proprietary assets, ownership intensity. There is a *public* interest in the *personal* ownership of land. The fourth criterion, liquidity or marketability, may be different for the public if there is 'subjective illiquidity'—in other words, if the owner refuses to sell even when there is a ready market because the land is a personal asset.[2]

Social responsibilities

In industry and national economic policy, 'paternalism' is often a pejorative term connoting interference by employer or state in matters which people best manage for themselves. Social responsibility can be an excuse for wasting someone else's money; the true social responsibility of the businessman is to make profits.

---

[1] The extreme example is the success of the Trust's Operation Neptune (which concerns the whole British coastline) and tree-planting appeals (which concern properties spread throughout the country.)

[2] Subjective illiquidity is the effect of emotional sunk capital which raises the value of an asset to its owner relatively to its value to others. Subjective illiquidity is in the public interest if it induces a traditional owner to continue his ownership of a heritage asset (Section V); and there can be a similar public interest in the continuation of a family firm. For passively held non-heritage assets, by contrast, the principal effect of subjective illiquidity is to raise their price to outsiders.

In the country, paternalism shows another face. The larger landowner has traditionally played a leading part in the life of the local community and has felt himself responsible for the welfare of tenants and employees, including the retired. And this attitude extends to keeping the land attractive to look at and live on or near. The economic basis of such responsibility is long-term planning and development, sound investment, and balanced land use, reconciling conflicts of interest and uniting a variety of professional skills.

The large private landlord typically reroofs buildings with traditional materials like thatch or slate, when corrugated asbestos would be cheaper, and refrains from cutting down trees and rooting out hedges even at a material cost in income forgone. Small woods can be maintained (and kept open to the public) which would be uneconomic for a body like the Forestry Commission. Many small woods would be lost under alternative systems of ownership. Forestry, farming and running the house and grounds are complementary activities which help to support or even cross-subsidise each other. The large private landlord typically tries to integrate these various activities by looking at all aspects of use in the round.

In a survey of 1,430 estates, Professor Denman found that social responsibility was one of the three main inducements to landownership, coming little behind residence and inheritance.[1]

The state and its agencies are unsuited to these functions, the essence of which is the landlord's expenditure of his own money on partly or wholly public causes. The traditional institutions, like the Oxford and Cambridge colleges, come nearest to the individual landowner since they share much of his outlook. The financial institutions say that they are trying to do the same. The Northfield Committee reported:

> 'The financial institutions told us that like the traditional institutions they intended to follow in the footsteps of the best private landlords and furnished us with examples of social and environmental expenditure for which there can be no financial return.'[2]

But there is a conflict of interest here; financial institutions are ultimately responsible to their investors, who do not live on the

---

[1] D. R. Denman, *Estate Capital: The Contribution of Landownership to Agricultural Finance*, George Allen and Unwin, London, 1957, p. 113 and Appendix.

[2] Northfield Report, *op. cit.*, para. 342-4.

land where their money is invested, and have not the same motive as a resident individual to spend money on financially unrewarding purposes.

> 'In general we accept what the institutions say about their approach to agricultural landownership and do not think that they are likely to be motivated solely by narrow financial interests. But we think that as new landlords of all types enter the market there is a danger of the desirable features of private landlordism being lost.'[1]

Interests of employees

The employee is interested in having a variety of prosperous potential employers to avoid dependency on the favour of any one individual or organisation. All forms of landownership satisfy this criterion except land nationalisation. Land nationalisation would be likely to create jobs, but mostly in offices. Small-scale owner-occupation can destroy jobs, since much of the work is done by the family. The widest variety of rural employment is provided by the larger estates in which, besides a small estate office, the works department may employ joiners, plumbers, masons, painters and other maintenance staff. There may also be a forestry department, a game department and a market garden. The variety and integration of land uses increase the variety of employment.

The feeling of social responsibility among larger landowners is especially strong towards staff and former staff. It has been traditional practice to look after retired employees, many of whom continue to live on the estate. In return, employees often give many years of service to a single estate and those who wish can live out their lives in the community where they were born. Similarly, although agents nearly always come from outside the local community, they generally develop a strong sense of identity with the estate and often spend several decades in the employ of the same landowner.

Interests of tenant farmers and new entrants to farming

The interest of tenant farmers is the opportunity for farming without the burden of the capital cost of a farm and with enjoyment of security of tenure at a market rent.[2] Present

---

[1] Northfield Report, *ibid.*

[2] For the meaning of 'security of tenure at a market rent', above, page 58, note 1.

security of agricultural tenancies goes far beyond this and can extend over two generations or more at rents well below the market-clearing price. The potential farmer, the new entrant to farming, has lost from this state of affairs as much as the landlord—or more. A special relationship between landlord and tenant is traditional in farming and rents have often been lower than the market would bear. But the manufactured rights of tenants now reach so far that landlords are reluctant to create new tenancies and are left with little alternative but to operate through partnerships or take the land in hand themselves.

This situation is particularly regrettable since the recent rapid rise in the price of farm land and in stock prices has made owner-occupation, the only alternative, more expensive and less practicable. The traditional farming ladder relied on the landowner to provide the capital value of the farm while the tenant provided the capital for the stock and enjoyed the opportunity to save for a farm of his own. The need for this ladder is stronger than ever among those tenant farmers who wish to become owner-occupiers; but its efficient working is impeded by the excessive security of agricultural tenancies.

The farming ladder was a natural evolution of the landlord/tenant system. The financial institutions are generally less willing than personal landowners to let to inexperienced farmers. In addition, the institutions are also affected by security of agricultural tenancies and have consequently become more reluctant to let to individuals, often preferring to let to companies or to take the land in hand themselves.

Interests of residential tenants

The responsible attitude towards staff extends also to agricultural[1] and residential tenants. There is evidence suggesting that tenants on a large estate prefer a personal landlord to any alternative.

In the West Country, where the Bedford estates then com-

---

[1] The Buccleuch estate, for example, granted its tenants lime subsidies in the 1920s—long before the subsidisation of lime became government policy. The dispersal of financial decisions through personal landownership has enabled progress to be made even in times of general economic stringency—for instance, through the financing of siloes. As an example of action by a personal landlord to offset a natural calamity, Eaton rents were reduced by the Grosvenor estates in 1968 to offset the effects of foot-and-mouth disease, and replacement cattle were brought in from Ireland.

prised about 26,500 acres in Devon, Cornwall and Dorset, a rumour of impending sale in 1910 caused tenants to draw up petitions to the Duke:

> 'We . . . having experienced your kindness and generosity as a Landlord, hear with great regret that you are contemplating selling your property. . . . we would much prefer continuing your tenants than being in any possible position resultant from sale. We feel sure that the contemplated action would be disastrous to Tavistock generally.'

But the requirements of taxation were inexorable and the property was sold.[1]

Legislative control of domestic rents has much the same consequences as security of agricultural tenancies: the sitting tenant gains at the expense of both the landlord and potential tenants who might be willing to pay a higher rent. The solution to both problems requires a rise in the rent to a market-clearing level.

Interests of neighbours

The interests of neighbours include a local community with a thriving economic and social life and land and buildings maintained in conformity with local styles and conditions. This involves respect for local architecture and materials and expenditure on hedges, woodlands and the like even where there is no or no adequate economic return.

The first three of the governing principles at the start of this section suggest that family ownership is the form of landownership likely to satisfy these criteria best. A family can live on an estate and have roots there; a state agency or a traditional or financial institution cannot. As a country landlord, a family's ambitions are generally confined to maintaining and improving the estate in its present pattern of uses. Thus neighbours have some safeguard against unwelcome change. And the family has both more opportunity and more incentive to take the long-term view, even at additional cost or at the expense of profit forgone, than has a financial institution with a duty to secure the best

---

[1] In addition to estate duty, taxes on this land included Increment Value Duty at 20 per cent, Reversion Duty at 10 per cent, Undeveloped Land Duty at 0·2 per cent a year and Mineral Rights Duty at 5 per cent of annual rental value, all four having been introduced in 1909. For a more recent example of a sale forced by estate duty, the sale of Iona by the Argyll estate is referred to below, pp. 89 and 95, note 3.

return for its investors.[1] The fiscal system need not be biased in favour of the personal landlord to secure these advantages; they will be secured by economic forces if the system is not biased *against* him, as at present.

The contribution personal landownership can make to the life of a community gives a new sense to the concept of a proprietary community. It is an idea that may not come naturally to the town-dweller; in a block of flats it is possible to live for years without knowing who most of one's neighbours are. Heavy capital taxes and the consequent difficulty or impossibility of handing wealth on, however undesirable the taxes may be, may do less damage in such surroundings than in the country.

'The rural landowner discharges many social functions', declared Archbishop Temple, 'and ownership of agricultural land, subject to consideration of the public welfare, should not be subject to the same restrictions on ownership as industrial stocks and shares.'[2]

Recreational interests

Recreational interests are interests in the use of land for other than such productive purposes as agriculture, forestry and fish-farming. They include shooting and fishing for sport, where a crop is harvested each year but the profit motive is subsidiary or absent. They thus cover many forms of pleasure and enjoyment in which most of the population are interested directly or indirectly.

The last of the five 'governing principles'[3] was that interest in the countryside is not confined to residents and visitors but extends to all who wish to preserve its beauty and resist its destruction. Similarly, the importance of an area or piece of land should not be judged primarily by the number of visitors it attracts. The Lake District is over-visited. There is a need for places where it is possible to be alone, like Hulne Park at

---

[1] 'I took a conscious decision that it was important to keep the villages alive. . . . It's totally uneconomic—we could make ten times as much out of the holiday trade—but I think it's the right thing to do. (Lord Coke discussing his Norfolk estate with Graham Turner, *Daily Telegraph*, 13 November 1981.)

[2] William Temple, *Christian Faith and Life*, Student Christian Movement Press, London, 1945, p. 47. An example of what a personal landowner can contribute to the life of the neighbourhood is the summer festival run for many years until her recent death by Lady Birley at Charleston Manor, Alfriston.

[3] Above, pp. 66-68.

Alnwick, where the public have access throughout the year provided they leave their cars outside.

'The countryside and the heritage' is the title of Section V. But there is a distinction only of degree between famous properties and the less well known. The rôle of the private owner is if anything even more important in caring for less well-known properties, because no one else will do this work if he does not. The public interest coincides with the private interest of the landowner because he is providing a public asset in the act of caring for his own property.

In the larger estates, the complementary relationships of the various departments have made it possible to provide amenities which for reasons of cost it would be difficult or impossible to provide otherwise. J. T. Coppock and B. S. Duffield have written:

> 'Until the beginning of the twentieth century, the characteristic pattern of rural landholding was that of an estate, divided into tenanted farms, but with the woodlands and usually a home farm kept in hand. . . . [These estates] had in common the fact that they were planned as a unit, with agriculture, forestry and often field sports and amenity considered from the viewpoint of the resources of the whole estate. With the break-up of estates and the rise of owner-occupation, which now accounts for more than half of all farmland, such a synoptic view of the land has become less common. . . . There is also evidence that owner-occupiers are less likely to take a broad, long-term view of the resources they control.'[1]

Interests of voters

Apart from any involvement in their personal capacities as landlord, tenant or employee, the voters have a twofold interest in country landownership. First, as consumers and taxpayers they are interested in the efficient and productive use of farmland and woodlands; this is like their interest in the economic use of any other resource, including labour. Secondly, as taxpayers and citizens they are interested in the countryside as a place for enjoyment and a heritage of natural riches which requires respect and care if it is not to deteriorate.

Both interests raise the question (to which we return in Section VII) whether the present taxation of landownership

---

[1] J. T. Coppock and B. S. Duffield, *Recreation in the Countryside: A Spatial Analysis*, Macmillan, London, 1975, p. 142.

serves the taxpaying community as a whole. Is the taxation of personal landownership so heavy as to restrict the efficient use of land and its availability for purposes of public enjoyment which other forms of landownership would serve less cheaply or not at all? In addition, the second of the two interests poses the usual problem of how far the 'externalities' can be 'internalised'. In other words, how far can the apparent divergence between private and public interests which the term 'externality' implies be reduced or eliminated by giving private persons a stronger motive for doing what the public interest requires?

Internalising externalities generally involves the creation or redistribution of property rights. The arguments of Section III and the present section have suggested that personal landownership is in general the form of country landholding most likely to maximise wealth externally as well as internally because it achieves a closer coincidence of private and public interests than can be achieved by landownership in other forms, especially ownership by government or its agencies.

A survey of public opinion

Are these conclusions acceptable to the general public?[1] 'There is still a feeling', says Douglas Sutherland, 'which is a direct heritage from the eighteenth and nineteenth centuries, that the landowners hold too much power and have too many rights.'[2] Thomas De Gregori writes of 'a large body of literature in economics and other diverse sources that treat property rights in land as being of a different kind from other forms of property rights'.[3] The 'other diverse sources' are mostly literary ones and the 'difference of kind' is to the disadvantage of land. Although a number of economists, including some cited in this *Hobart Paper*, have been sympathetic towards landowners and have emphasised their community of interest with society at large, it is probably true that the literary and political

---

[1] The question is not meant to imply that what is alleged to be politically acceptable is a reliable guide to sound policy or even to the realities of political acceptability itself. (W. H. Hutt, *Politically Impossible . . .?*, Hobart Paperback 1, Institute of Economic Affairs, 1971.)

[2] Douglas Sutherland, *The Landowners*, Anthony Blond, London, 1968, p. 158.

[3] Thomas R. De Gregori, 'In Perpetuity: Some Reflections on Literary Views of Land and Other Forms of Property', *American Journal of Economics and Sociology*, July 1979, p. 234.

treatment of landowners has for many years been more hostile than friendly.[1] Professor Hayek has pointed out how literary treatment of such themes can influence the opinions of the general public.[2]

There is little direct evidence of public opinion towards different systems of landholding. But a survey commissioned in 1978 by the Country Landowners' Association from the British Market Research Bureau suggests that the general public are not unaware of the advantages of personal landownership. More than 2,000 people from all parts of England, Scotland and Wales were asked their opinions on various propositions about landownership on a six-point scale, the six possible answers totalling 100 per cent: 'agree a lot'; 'agree a little'; 'neither agree nor disagree'; 'disagree a little'; 'disagree a lot'; 'don't know'. In answer to the proposition 'Agricultural land should be acquired by the State', 73 per cent expressed disagreement, including 61 per cent who disagreed strongly ('disagree a lot'). In answer to 'It is important for a free society that agricultural land should be privately owned', 70 per cent agreed, including 56 per cent who agreed strongly. In answer to 'The traditional links between families who own land and the places where they live are good for the countryside and those who live there', 76 per cent agreed, including 52 per cent who agreed strongly. In answer to 'Owners of agricultural land perform a worthwhile service to the community', 86 per cent agreed, including 65 per cent who agreed strongly. If private agricultural land were to be nationalised, 65 per cent thought the beauty of the countryside would be less well cared for and 64 per cent thought that those who would manage the land would understand the needs of the countryside and its inhabitants less well than the present owners.

---

[1] On a related theme, Michael Jefferson has described the hostility towards industrialisation in the 'Condition of England' novels of the 1840s and 1850s. ('Industrialisation and Poverty: In Fact and Fiction', in *The Long Debate on Poverty*, IEA Readings No. 9, Institute of Economic Affairs, 1972, 2nd Edition, 1974.)

[2] F. A. Hayek, 'History and Politics', in his *Studies in Philosophy, Politics and Economics*, Routledge and Kegan Paul, London and Chicago, 1967.

V. THE COUNTRYSIDE AND THE HERITAGE

Real and factitious difficulties of conservation

Land is at the centre of the effort to conserve the heritage—not only farmland, wasteland and forests but town and country estates and the collections they contain.

Some of the difficulties of conserving the heritage are real—expensive, or even impossible to overcome. There are attacks by the forces of nature: it is a losing battle to try to preserve the cliffs below Beachy Head. There are attacks by Mammon: it is costly to prevent the National Coal Board from mining the Vale of Belvoir. There is natural wastage in the defending forces: the struggle to maintain the large churches of Lavenham, Melford, Sudbury and elsewhere in Norfolk and Suffolk is primarily due to the movement of population away from these clothmaking towns since the days when their churches were built.

Other difficulties, while serious for the owners and their collaborators, need not be so for the country as a whole. In particular, all historic buildings were erected at a time when the cost of labour was much lower than now and they were designed accordingly. But, since the increase in labour costs is largely a reflection of a rise in the general standard of living, there should be more resources, not fewer, to keep up heritage assets. This result might not follow if conservation were an unpopular cause; but, on the contrary, it has been one of the most fashionable, and increasingly fashionable, causes of the last decade, on which much time and money have been spent by an army of volunteers.

Most of the difficulties in conserving the heritage are not real or natural but man-made. Their core has been the attack on the heritage through taxation by governments of both political parties.

Friends and enemies of conservation

Conservation has enemies as well as friends. Destruction has more than one motive and more than one meaning. Motives range from pure greed (as in the destruction of the Inca treasures[1]) to the wish to damage an adversary; and the adversary may be personal or impersonal, military, religious, political, cultural, or social.[2]

However brutal the work of destruction, it has had its theoretical supporters if its motives lay within the realm of ideas. The work of Leo the Iconoclast was sanctified by the highest and most formal ecclesiastical authority.[3] What this attitude has meant in England is illustrated in detail by a document called simply 'Inventarium Monumentorum Superstitionis' which lists the contents of 150 Lincolnshire churches destroyed as superstitious or unnecessary in 1566 and specifies the methods of their destruction.[4] Examples from our own time include the cultural conflict between old and new. Islington and Camden Councils, among others, have torn down streets of pleasant old houses, cheap and easy to repair, in order to replace them—at many times the cost—with council houses reflecting different cultural values.[5]

A more subtle alternative to the bulldozer is the use of the adversary's buildings for perverse, ironical or insulting purposes. 'The French revolutionists', as Ruskin called them, made stables of the cathedrals of France. John Cook in 1793

---

[1] William Prescott in *History of the Conquest of Peru* gives no suggestion of any religious or even political motive (Book III, Chs. V-VII).

[2] There seems to be a place for an economic theory of destruction within the expanding ambit of topics to which the methods of economic analysis have recently been applied for the first time (charity, altruism, marriage, riots and revolution, tax avoidance and evasion, etc.). Professor George Stigler of Chicago has described this process as a form of economic imperialism. Destruction has value for the agents even though it generally inflicts bigger losses on others. Analysis may assist its exorcism.

[3] 'After a serious deliberation of six months', says Gibbon of the Synod of Constantinople in 754, 'the three hundred and thirty-eight bishops pronounced and subscribed an unanimous decree . . . that image-worship was a corruption of Christianity, and a renewal of Paganism; that all such monuments of idolatry should be broken or erased . . .' (*The Decline and Fall of the Roman Empire*, Chapter XLIX.) It is noteworthy that Gibbon's sympathies were entirely with the iconoclasts.

[4] Reproduced in Edward Peacock, *English Church Furniture, Ornaments and Decoration, at the Period of the Reformation*, John Camden Hotten, London, 1866. The destruction of buildings and works of art in Communist China during the 'cultural revolution' and in Iran since the overthrow of the Shah constitutes similar behaviour.

[5] More finely balanced judgements are involved in decisions whether to restore a building to its original state: for example, whether to strip a medieval church of its 18th-century decoration.

wanted to do the same with the chapel of King's College, Cambridge.[1] The Soviet revolutionaries turned churches into anti-God museums. British republicans have proposed that Buckingham Palace be turned into council flats.

Conservation, by contrast, respects the spirit as well as the fabric. It requires that buildings should be used as far as possible for the purposes for which they were constructed: churches as churches, not concert halls; houses as houses, not museums or offices or hotels. If this aim is unattainable, the new use should at least be as near as possible to the old.[2]

An important example of this principle is that long association between a house and a family is one of the assets which it is the business of conservation to conserve. It is part of the history of the country. A new family is better than no personal owner but not as good as the old. The original family may be no less important than the original furniture. Long associations are irreplaceable assets, like Old Masters, and very long associations are in very limited supply, like rare Old Masters. The loss to the country from the sale of Stonor Park and the dispersal of its contents was all the heavier because the house had been in the possession of the Camoys family for over 800 years, perhaps the longest unbroken occupation of any house in Britain.

These commonsense values have been under attack for the last century or more from egalitarian taxation, the central thrust of which has been to attack personal benefit from the ownership and use of resources while offering a measure of relief in so far as the personal link is attenuated. Whatever the merits of this policy for taxing a group of people with varying wealth but similar life styles, it is completely unsuitable for taxing owners of heritage assets. Indeed, it is so constructed as to do the maximum damage to the heritage for every million pounds of revenue because it concentrates its attack on the immaterial part of the heritage which gives life to the rest.

---

[1] 'I wish to see all the churches down, and the roads mended with them, and King's Chapel made a stable of.' (Cited in *King's College Chapel: Comments and Opinions*, edited by Kenneth Harrison, 1956 edition.)

[2] If it is impossible, or prohibitively expensive, to keep buildings in their original uses, it is important they should not simply be allowed to deteriorate as a result of zoning and planning restrictions. Many of the best London buildings, such as those in Belgravia, Regents Park and Bedford Square, could probably not have been supported economically without a change of use from private residences to offices, academies and so on.

The tax system that produces such results may not do so entirely by design; as John Cornforth points out, the landed interest 'is often penalised by legislation intended to cover other aspects of economic life'.[1] But it seems more realistic to recognise that the fiscal attack on the personal link with heritage assets is at least partly intentional and that the loss it causes to the owners and others is desired as representing the defeat of an adversary or the victory of one culture over another.

This attitude cannot be shown to be mistaken, just as it is not demonstrably wrong to break images or stained glass. But analysis can help to identify or even quantify the net costs it imposes on society—the excess of the losses to the owners and others over the value of destruction to its agents. This is the theme of the present section.

Vocational ownership

Among the friends of conservation, the most committed are the personal owners. They have reason to be. Most heritage assets are *personal assets* and thus more valuable to the owner than to anyone else.

It is not only that 'these owners could more properly be called stewards or trustees'[2] than beneficial owners (and some owners do regard themselves as trustees). Stewards are paid. Even trustees are not normally expected to contribute more than the value of their unpaid work. But the owner of a substantial town or country estate who feels he has been entrusted with a living entity embodying part of the nation's history, art or countryside lies under a heavier obligation which is not the less real for being self-imposed. His spendable income is reduced, not increased, by his ownership of heritage assets because of the requirement of funds for their upkeep. Individual assets are not available for sale, even though a market for them exists, because of the need to preserve the complementary relationship between the different parts of the estate if it is to remain a living entity. For the owner, the preservation of the heritage—his and the nation's—is an absorbing pre-occupation.

---

[1] John Cornforth, *Country Houses in Britain—Can They Survive?*, Country Life, 1974, p. 20.

[2] Patrick Cormack, *Heritage in Danger*, New English Library, 1976, p. 29.

Like the Stayed Man in John Earle's *Microcosmographie*,[1] he is 'not scatter'd into many pieces of businesses, but that one course he takes, he goes thorough with'.

The four values of an estate
Heritage estates have different values for different purposes; four different values are relevant to our argument. They are, from the highest down:

   (i) social value in current use;
   (ii) personal value in current use;
   (iii) break-up value;
   (iv) private value in current use.

(i) *Social value in current use* is the value to society of keeping an estate in its current use, including any beneficial effects on third parties, by comparison with the best alternative.

(ii) *Personal value in current use* is the satisfaction to the owner of keeping the estate in its current use (*satisfaction* in the sense of the economic term *utility*). It excludes third-party effects[2] except where they are reflected in satisfaction to the owner. Social value in current use can be *seen* to exceed personal value in current use whenever conservation bodies or others try to persuade an owner to carry on in his present position, or regret his unwillingness or inability to do so, or try but fail to find a successor. And social value in current use can be *expected* to exceed personal value in current use in more stable situations where this relationship cannot be demonstrated.

(iii) The *break-up value* is what the owner can realise by selling and pulling out of the area. Break-up can be partial or total. Since the owner is a free agent,[3] personal value in current use is by definition higher than break-up value. But it may be little higher, and it hovers uneasily above break-up value when

---
[1] Robert Allot, London, 1618.
[2] Effects on third parties are discussed in Section IV. These are the *external effects* or *externalities* of economic jargon. The net result of third-party effects is assumed to be positive (positive effects outweigh negative). In other words, the public gains rather than loses, or gains more than it loses, from personal ownership of heritage estates.
[3] It is significant that the hereditary owner is a free agent but not a volunteer, since he has never volunteered. A volunteer would be less likely to accept a large shortfall in the private value in current use below the personal value in current use.

the owner is losing heart, ceasing to care and on the verge of giving up.

(iv) *Private value in current use* is the value of the estate to the owner excluding all elements of vocation and service to the community. Since the owner is not on the breadline, it is not negligible and may even be substantial. But it may be far below the break-up value if the owner is making large personal sacrifices to soldier on. And it may be well below the wealth of someone in modest but comfortable circumstances with no responsibilities for heritage assets, especially when the owner is near the point where it is simply not possible, physically or financially, to carry on any longer.[1]

Vocational ownership may be defined as a situation in which *personal value in current use* exceeds *private value in current use*. It is a general concept and is not confined to heritage assets. The purchase of National Savings on financially unattractive terms in wartime, for example, is vocational saving, and their retention is vocational ownership. But *commitment* to heritage assets is the most striking and perhaps the most important example.

The tax system seeks in principle to levy capital taxes on or near the break-up value of heritage assets and thus to include the element of vocational ownership constituted by the excess of break-up value over private value in current use. The full rigour of this principle may be mitigated by reliefs. But the reliefs may do little to bridge the wide gap between break-up value and private value in current use, which latter is the true measure of taxable capacity. This has three unfortunate effects. First, it is unfair between taxpayers; assets with the higher break-up value may have the lower private value in current use, so that the heavier tax burden falls on the taxpayer with the smaller stock of assets for his private use. Secondly, tax on the excess of break-up value over private value in current use

---

[1] If an owner wishes to demolish, the *break-up value* in the economic sense is the break-up value in the contractor's sense and will be small or even negative if the value of the site does not cover the cost of demolition. The private value in current use is still smaller than the break-up value (or a larger negative). The social value in current use ought to be substantial if it is to justify a public authority in seeking to thwart the wishes of an owner by means of a conservation order, perhaps at the cost of a protracted legal wrangle. This example illustrates how the social value in current use can be far in excess (and a large multiple) of the break-up value and thus also of the personal value in current use (which may be little higher than the break-up value) and of the private value in current use (which may be much lower).

is a pure tax on vocation,[1] and this is undesirable in itself. Thirdly, the taxation of break-up value as though it were private value in current use gradually and systematically destroys the public interest in the personal ownership of heritage assets. In other words, taxing the committed owner on his commitment gradually and systematically destroys the public good represented by the excess of social value in current use over break-up value.[2]

As I have argued elsewhere, the only taxes that are ultimately unavoidable (apart from poll taxes and taxes on necessaries) are those on vocation in earning, saving and ownership.[3] But taxes on vocation are unavoidable only as long as the vocation is not eroded financially or by loss of morale. The list of great houses and their collections destroyed and dispersed in recent years[4] gives some idea of the losses inflicted on society

---

[1] A tax on vocation is a tax on altruism in earning, saving or owning. An example is the income tax levied on the sacrificial marginal earnings of a doctor who works during an epidemic until he is almost asleep on his feet. But for his sense of duty, he would prefer a good night's sleep to his marginal earnings even gross of tax, let alone net. Even gross of tax, he would prefer marginal leisure to marginal income. The levying of tax on the marginal income must *either* diminish his supply of services *or* constitute a pure tax on his altruism *or* combine both these effects in varying degrees. Any of these results is either unjust or inefficient or both. Similar arguments apply to vocation and commitment in saving and owning.

[2] This is the *external social value in current use* (below, page 85, note 1).

If the situation of the vocational owner is compared with that of the vocational doctor, the social value in current use includes all the third-party benefits in both cases. For the doctor working in an epidemic, the private value in current use represents the preferred position exclusive of altruism and the personal value in current use represents the preferred position inclusive of altruism. The tax on marginal earnings is a pure tax on altruism and reduces the personal value in current use towards the break-up value. If personal value in current use falls below the break-up value, the tax on marginal earnings is a total bar to altruistic endeavour and the doctor reverts to the egoistic position at which private value in current use increases to the break-up value. Similarly for the vocational owner. His commitment is the excess of personal over private value in current use. The heavier the tax burden, the smaller the excess of personal value in current use over the break-up value relatively to the excess of the break-up value over the private value in current use, and the stronger the incentive to economise on altruism by realising the excess of the break-up value over the private value in current use. The scales are finally tipped when taxation pushes the personal value in current use below the break-up value.

[3] *Tax Avoidance and Evasion: The Individual and Society*, Panopticum Press, Upminster, 1979, through The Alternative Bookshop, 40 Floral Street, London WC2, p. 100.

[4] A partial list up to 1974 is provided by Cornforth in *Country Houses in Britain*, *op. cit.*, pp. 5-11.

by the taxation of heritage assets at sums far in excess of their private value in current use. These social losses may from the beginning, and must in the end, far exceed the losses to the owners. The loss to society does not diminish over time.

The public interest in vocational ownership

Not all vocation is in the public interest, even if the public interest can be identified and agreed. In public life, the worst damage is often done by people acting out of a sense of vocation. In private business, classical liberals have believed since the time of Adam Smith that a man generally serves his fellows best by pursuing his own interests.

There is nevertheless a place for altruism within a market economy. Indeed, the market economy and the personal ownership on which it rests are the most efficient means of satisfying altruism, just as they are the most efficient means of satisfying egoism.[1]

The public interest in vocational ownership is threefold. First, there is economy in running costs; property costs less to look after when it is owned by an individual or a corporate body with a strong sense of identity, like a school or a monastery, than when responsibility is spread widely or divided among people with disparate or conflicting interests.[2] Secondly, there is the intangible public interest in personal ownership; personal ownership of heritage assets is a *public good*.[3] Thirdly, it is

---

[1] The economic theory of uncoerced altruism is a neglected subject. Pioneering work has been done by Professor Gary Becker, summarised in his lecture 'Altruism and Selfishness in the Family and the Market Place' (London School of Economics, 20 May 1980). Professor David Collard in *Altruism and Economy* (Martin Robertson, Oxford, 1978) uses the term *altruism* in a different sense involving government coercion to transfer the financial costs of 'altruism' to third parties.

[2] Democratic management of property seems to be costly and inefficient whether the democracy is nominal or real. When property is owned by central or local government, politicians and officials are seldom committed to its long-term welfare and democratic control is only nominal because electors generally have little influence on what is going on. When property is in the hands of a London club, by contrast, its upkeep may be inefficient because democratic involvement is excessive and too many members wish to have their say on day-to-day matters beyond their expertise, such as the repair of the boilers. This advantage of personal ownership is a form of income-efficiency, although like other annual sums it has in principle a capital equivalent.

[3] A *public good* in the economic sense is a good or service the use of which by one citizen does not diminish the supply available to another; military defence and

[*Contd. on p. 85*]

often forgotten that the public interest includes the interest of the owner.[1]

Extent and direction of the public interest in vocational ownership

The public interest in the personal ownership of heritage assets may much exceed the value of the same assets to the owners in their private capacities—the private value in current use. The *social value in current use* may be several times as large as the break-up value and the *private value in current use* several times as small, even if the *personal value in current use* exceeds the break-up value because heritage assets are personal assets.

The public interest in the personal ownership of heritage assets could perhaps be measured along the lines of the cost-benefit analysis used in recent years to quantify the benefits to the community of expenditure by government, even though many of the critical assumptions on which these calculations are based are inevitably arbitrary and debatable within wide limits.

But more important than measurement is direction. Where the argument is leading is more important than how far it goes or how fast. The general desirability of private ownership of heritage assets is established by the various theoretical and

[*Contd. from p. 84*]

street lighting are stock examples. Private ownership of heritage assets is a public good in the economic sense: the social value in current use exceeds the break-up value, and this excess need not be diminished by an increase in the number of citizens to whom it yields satisfaction. The private ownership of heritage assets is also a *costless public good* in the sense of being an advantage to society while imposing no economic cost except on the owner, because the economic cost is borne by the shortfall of private value in current use below the break-up value. The private ownership of heritage assets is perhaps the only public good that is completely costless in this sense.

[1] Social value in current use, being the highest of the four values mentioned under 'The four values of an estate' (page 81), must by definition include all these three elements of the public interest in vocational ownership. Break-up value excludes the second. It also ignores the first, which is included in both personal and private value in current use. Break-up value introduces a second dimension, since the capital value of any additional income-efficiency of personal ownership is in abeyance in so far as the tax system is biased against personal ownership and in favour of government ownership.

Social value in current use may be divided into two components: (i) the break-up value (which can be acquired on commercial terms if the owner sells) and (ii) the *external social value in current use*, which is the excess of the social value in current use over the break-up value. It is the *external social value in current use* which is a *public good*.

empirical arguments set out earlier, which merely confirm what is suggested by common observation.[1]

The case to the contrary is both less immediate and less certain of direction. It is concerned not with day-to-day affairs but with statistical abstractions like the inequality of wealth which have no obvious interpretation or significance. It is difficult to establish a public interest in any particular change in the inequality of wealth (or even in its direction). And arguments from inequality are uncertain in direction in the sense that the same policies that reduce the inequality of wealth may (and in the British context very likely do) increase the inequality of spending, that is, the gap between the living standards of richer and poorer.[2]

Ingredients and evidence

What are the elements which make up the intangible public interest in personal ownership; and how can they be shown to be real?

Under 'Principles governing the public interest in landholding systems' (page 66), five were mentioned, of which the first three apply at least as strongly to heritage assets as to others:[3] in the country resident landlords are generally preferred to absentees; most people in their capacity as residents believe all change is for the worse; and a financially less attractive long-term plan may be preferable on other grounds to a financially more attractive alternative. Government has been trying to do the landlord's job, not because it is more suited to the task (rather the opposite), but because it has been taxing its competitor out of existence. 'In seeking to influence the broad environmental questions of good husbandry, soil preservation, pollution control, livestock management and cropping practice', says Clifford Selly, 'the Government is in effect

---

[1] The few apparent exceptions serve to strengthen this general argument. In particular, the Ministry of Works (now the Department of the Environment) has proved an admirable custodian of numerous ruins. But there is a public interest in fending off the use of this argument for the government ownership of heritage assets which are at present in working order.

[2] This argument is spelt out in *Is Capital Taxation Fair?*, *op. cit.*, pp. 61-67. The general argument is well illustrated by the distinction between break-up value and private value in current use and by the potentially large absolute and relative excess of the former over the latter.

[3] The fourth and fifth are discussed later, respectively under the headings 'Qualifications' and 'Conservation bodies and pressure groups'.

assuming the mantle of the landlords of old. It is really engaging in estate management on a national scale.'¹ The arguments in favour of personal ownership and against government's assumption of the mantle are even stronger for heritage assets, not least because personal ownership retains individuality of taste and style.

The public interest in heritage assets is conservative, as the concept of conservation implies. Churches are best left as churches, homes as homes, preferably for the traditional owners. Alternative uses are at best a second-best. And the whole is more than the sum of the parts.² We return later to the idea of the whole and the parts under the heading 'A going concern'.

A community, like a business, can succeed or fail as a going concern. Indeed more so, since a community is an organic growth whereas a business can be created by conscious design. You cannot make friends and neighbours as you can make cakes; witness the failure of the post-war new towns to create natural communities. Damage to a community is in this sense irreversible. The damage of war may be made good, as in De Halle at Ieper, only a few stones of which remained above ground in 1918. The damage of nature may be made good, as in the cathedral of Messina, faithfully reconstructed after every earthquake. But a community damaged or destroyed is damaged or destroyed for ever, and even its revival is an act of providence similar to its original creation.³

The value of personal ownership and the commitment it implies is none the less real for being intangible. The damage

---

[1] *Ill Fares the Land: Food, Farming and the Countryside*, André Deutsch, London, 1972, p. 138. Among the reasons for the unsuitability of government for this rôle, especially on a national scale, is Denman's 'first law of proprietary magnitudes' (cited on page 56) exemplified in the comparative cost figures on page 57.

[2] 'Where a house still contains fine contents or collections surrounded by gardens, grounds or park making it an indivisible whole, the solution must be for the family to remain in the house and open it to the public. The houses in this booklet have virtually all of them lost their contents, and it is this sad fact that makes them ideal candidates for alternative use.' (Sophie Andreae and Marcus Binney, *Tomorrow's Ruins?—Country Houses at Risk*, SAVE Britain's Heritage, London, 1978, p. 2.) A list of houses whose contents have been lost in recent years, mostly as a result of taxation, is given in Cornforth's *Country Houses in Britain*, op. cit., pp. 10-11. The corresponding loss of 'gardens, grounds or park' is not the less real for being more gradual and less easy to pinpoint.

[3] The same argument applies to an institution like a school or a club as much as to a geographical community.

done by government through taxation is none the less real for being immeasurable. These general principles are supported by a variety of evidence, as the following examples show.

High cost of state ownership and maintenance

The idea that additional taxation of heritage assets in private hands would cost the government money was not only the common thread linking almost all the submissions to the House of Commons Select Committee on a Wealth Tax but was also accepted by a large majority of the committee (including MPs of both main parties) as the basis of their deliberations.[1] This argument combines the general proposition that the fisc can be too greedy for its own good[2] with the particular consideration that the personal owner of assets is their most economical steward.[3]

The Forestry Commission is unwilling to take on plantations below a minimum of some 100-200 acres. The costs are prohibitive even for a government body. But 100-200 acres is not a sound barrier beyond which everything goes into reverse. The reasons why it is more efficient and economical for small plantations to be in personal rather than government ownership also hold good for large plantations. The difference is that the comparative advantage the personal owner derives from local knowledge and involvement gradually diminishes and there is a rapid increase in the competitive advantage the state derives from subjecting its private rivals to prohibitive income and capital taxes from which it is itself immune.

As for the material ingredients of the personal ownership of heritage assets, likewise for the immaterial. The evidence supports the argument that there is a public interest in personal ownership of heritage assets. The death of an owner, or his inability or unwillingness any longer to bear the responsibility of ownership, causes unwelcome changes, perhaps a crisis,

---

[1] *Select Committee on a Wealth Tax, Session 1974-75:* Vol. I (Report and Proceedings of the Committee), HC 696-I; Vol. III (Minutes of Evidence, Sub-Committee B), HC 696-III, HMSO, 1975.

[2] All taxes have a point beyond which the yield falls as the rate of tax rises; the tax starts to become prohibitive and the base diminishes more rapidly than the rate of tax rises.

[3] 'It is worth underlining the costs of state ownership and, conversely, the purely financial benefits of inter-dependence of private and institutional owners and the nation.' (*Country Houses in Britain—Can They Survive?*, *op. cit.*, p. 122. Cornforth argues the point in some detail.)

locally or even nationally. Mentmore is only one of the more recent and notorious examples.[1] Most of the losses of houses and their contents listed by Cornforth in *Country Houses in Britain* are due to the weakening or ending of a personal link with the property, generally as a result of taxation. When agents prepared to put Iona on the market in 1979 to help meet estate duty liabilities on the death of the 10th Duke of Argyll, there was widespread local and national concern. It was reported that the Mormons were interested in buying the island. The National Trust for Scotland initially thought the cost of purchase was beyond their resources. There was a project for raising the funds from the individual members of the Church of Scotland. An offer from the Hugh Fraser Foundation was accepted in principle by the Duke's trustees. In the end the Foundation's money was used to finance the purchase of the island by the National Trust for Scotland.

Although this was perhaps the happiest outcome in the circumstances, no one gains from transfer taxes on heritage assets. They create unnecessary crises and emergencies. They disturb a generally acceptable situation and substitute a generally less acceptable alternative. They replace personal capital, which is the cheapest and most efficient means of financing heritage assets,[2] either with private funds diverted by the crisis from more suitable employments which would otherwise have had priority, or with government funds which represent the dearest and most inefficient source of finance because personal satisfaction and involvement are reduced to a minimum. The revenue from capital transfer tax on heritage assets is probably much less than the direct and indirect cost of the increased government responsibility for their upkeep imposed by the tax itself, let alone the loss to the public of buildings destroyed and their contents dispersed.

While a house remains in personal ownership it offers the possibility of recruiting voluntary labour to assist in its upkeep.

---

[1] It is ironical that the Mentmore fiasco, one of the largest losses of heritage assets in British history, was precipitated by the estate duty charge on the death of the 6th Earl of Rosebery in 1974, for death duties were introduced by Sir William Harcourt in 1894 when the 5th Earl of Rosebery was Prime Minister.

[2] Personal capital is the *cheapest* means of financing heritage assets because it is subsidised by the pleasure of personal ownership. It is the most *efficient* means because the 'subsidy' is economically costless in the sense that the pleasure of personal ownership has no alternative use and disappears if assets are no longer owned personally.

Lady Salisbury has assembled a team of skilled assistants who work without payment to help maintain the fabrics and tapestries. This is the 'external' counterpart of the 'internal' pleasure of ownership (the pleasure of the owner himself). Few people would do such work for nothing if the house were owned by a museum, a financial institution or the state. Voluntary workers work for voluntary organisations or for individuals.[1]

The preservation of a family home *as a home* in the hands of the same family is a good cause capable of attracting supporters who will vote with their hands as well as their feet. The reason is that it satisfies the central principle of conservation, that buildings of quality and interest should continue to be used wherever possible for the purposes for which they were designed and for which they have been used in recent centuries.[2] An individual willing to take on a large and expensive property which would otherwise deteriorate is performing an important public service.[3] But this work, however meritorious, is not an adequate substitute for the public interest in the continuing occupation of a family home by the same family.

The thrust of tax policy during the present century has been increasingly hostile to the personal and family use of heritage assets. It has been exactly the opposite of what the public interest requires. It is paradoxical that the Duke of Devonshire should for tax reasons need to be a *tenant* of the part of his family home remaining in family use, though it is the natural consequence of existing taxes. The public interest in Chatsworth is the same as the family's, namely that it should remain as far as possible a family home. By contrast, though in the general public interest, Lady Salisbury's policy of running Hatfield House as a single whole—and not as two separate

---

[1] The voluntary organisation which has perhaps had most success in attracting voluntary labour for the upkeep of the heritage is The National Trust, which has always preferred properties to remain in personal ownership where possible and has always done its best to preserve the links between its own properties and their former owners.

[2] This principle of conservation is reflected in the commitment to a building represented by voluntary donation in cash or in kind. If fiscal and other government interference is unimportant and the fate of a building is decided by the market, it remains in its traditional use as long as the satisfaction or utility that this use yields to its effective supporters outweighs the opportunity cost of forgoing a commercially more profitable alternative.

[3] A recent example is the American, Mr Robert Parsons, who has rescued Newark Park, Wotton-under-Edge, Gloucestershire.

entities, the family's wing and the rest—runs counter to the recent trends of tax policy.[1]

The public interest in the involvement of a personal proprietor in his estate is a principle that holds good for urban as well as rural property (Section VI) and for the various third parties mentioned in Section IV. The third-party interests are summed up in the concept of 'a going concern', analysed later.

Qualifications

It may be accepted that personal ownership of heritage assets is in the public interest (even if the interests of the proprietors are excluded from the public interest). But does this conclusion hold universally and without qualification?

The *positional economy* of Hirsch is concerned with assets subject to absolute or socially imposed scarcity or to congestion or crowding through more extensive use.[2] But, whatever the insights conferred by this analysis into competition for the best suburban houses and the best jobs in the civil service, it has nothing to do with heritage assets whose break-up value exceeds their private value in current use.

Likewise, and more generally, for the emotion of envy which has supplied the explicit or implicit foundation for so much of welfare economics. There must be very few who covet the task of looking after a large historic house at their own expense; indeed, anyone who assumes this burden gratuitously is rightly and warmly welcomed.

Conservation bodies and pressure groups: rôle of third-party interests

Although there is a substantial *public* interest in the *personal* ownership of heritage assets, it is not an interest easily translated into effective political pressure. This is partly because much of the benefit is spread widely and thinly and partly because the advantages are mostly long-term or intangible or both.

[1] Running a historic family home as a single entity is in the public interest excluding the interest of the family. But it may impose a cost on the family outweighing the benefit to others such that the interest of the rest of the public cannot be realised, even though that interest coincides with what would be the interest of the family were it not for the distortions due to taxation. A number of other estates, such as Buccleuch, also follow the policy of running a historic family home as a single entity.

[2] Hirsch, *op. cit.*

Evidence of the wide dispersal of third-party benefits which are large in aggregate is provided by the introduction of entry charges by some of our most visited cathedrals because visitors' donations were far from covering the costs of upkeep. Even large public benefits may be difficult to recapture by voluntary subscription and difficult to support through popular political activity.[1]

The unofficial conservation bodies, being themselves voluntary organisations, have their own reasons for saying little or nothing about the relative merits of personal and other forms of ownership. It is a question with political overtones, and they might lose members if they became involved. Of the 10 voluntary conservation bodies I approached on this subject, only four had something to say. The Ramblers' Association expressed a preference for public ownership of one type of asset in specified circumstances.[2] Hermione Hobhouse, Secretary of the Victorian Society, was inclined to agree that a substantial case could be made for saying that, in London, the best of the great estates provide considerable protection for townscape,[3] but she emphasised that this was a personal view. The Secretary of the Ancient Monuments Society pointed out that 'there are good and bad private and public owners of listed buildings', but went on to say:

> 'I do endorse the general view that private ownership produces a greater sense of pride than does public ownership. There is nothing like the fact of ancestral possession to create the imperative to preserve. A disproportionate number of derelict listed buildings are owned by local authorities.'[4]

[1] A number of houses open to the public, however, solicit moral support from visitors, many thousands of whom have signed statements expressing a preference for the retention of the house in its existing ownership and use. There seems little doubt that this represents the opinion of the large majority of visitors.

[2] 'With one exception, we have never expressed any preference for one form of ownership against another . . . The only exception concerns land in National Parks, where we have gone on record as expressing a preference for seeing more land brought into public ownership where this would have the effect of protecting outstanding landscapes against adverse forms of development, or of protecting public rights of access to open country.'

[3] The urban heritage is the subject of Section VI.

[4] He concluded: 'I do feel that there is a lot to be said for both private and communal ownership.' Particular examples of the latter are the success of the National Trust and the communal pride of a village or town in a historic church. And 'no private individual could do as well as does the Ancient Monuments Secretariat of the DoE in caring for ruins'.

The voluntary body with much the most explicit policy on the relative merits of personal and other forms of ownership, as well as the policy most favourable to personal ownership, is the National Trust.

'The Trust has no wish to acquire houses that are not in danger, believing that the best owner of a country house is generally the private owner. No organisation, however flexible or sensitive, can extend the same affectionate care to a house and its contents as the family who may have lived there for generations. The Trust regards its ownership as a solution only for special circumstances. ... The Trust prefers that its houses should be lived in ... [and] has no wish to create museums in the countryside, and over the years countless visitors have expressed their appreciation of the sense of continuing purpose and enhanced interest that occupation by the family of the donor confers.'[1]

'We believe that historic houses should whenever possible continue to be lived in as family houses, preferably by the family traditionally associated with the house and retaining the historic contents.'[2]

These statements are not only explicit assertions of precisely the principles I am arguing on economic grounds; they are also interesting for their wider social significance and implications. Of the voluntary conservation bodies, the National Trust is both the biggest (with nearly a million members) and the one most closely involved with government (through its unique statutory position in the capital taxation of heritage assets). And it is the heritage organisation which enunciated the strongest and most specific statement on personal ownership. This suggests that the principle of a public interest in personal ownership is widely accepted and supported and that the

---

[1] *The National Trust and the Country House*, National Trust, January 1979, p. 2. This is just the opposite of the East European policy of divorcing historic buildings from their former employments even when they are maintained to a high standard. Official British policy vacillates schizophrenically between these extremes. The logic of my argument is that in a non-communist country like Britain there is a public interest in the personal ownership of heritage assets. 'There is something contrived and clinical in all this rebuilding', says Cormack of post-war repair work in Warsaw and Leningrad, 'but the results have been astonishing and ought to act as a spur to those of us in the West who have so many genuine old buildings that merely demand our careful attention if they are to be safeguarded and appreciated into the next century. We also have the added incentive of being able to restore buildings that are still being used for their original purpose.' (*Heritage in Danger*, op. cit., p. 109.)

[2] Letter to the author from the Director-General of the National Trust, J. D. Boles, 4 February 1980.

National Trust has the self-confidence to articulate it because of its large membership.[1]

The idea that personal ownership of personal assets creates wealth for third parties as well as for the owner stands in contrast to the opinions of the National Union of Agricultural and Allied Workers (NUAAW) and the Trades Union Congress (TUC) on the related subject of the ownership of agricultural land. The NUAAW favours land nationalisation with compensation:

> 'One reasonable approach would be for former landowners to receive annuities for as long as they live, and until their children reach school-leaving age.'[2]

The TUC supports the joint proposal of the NUAAW and the National Executive Committee of the Labour Party that

> 'working farmers should be able to sell land to the state in lieu of capital taxation and then lease it back so as to maintain the farm business intact'.[3]

Whatever the case for these proposals on other grounds, they are precisely calculated to destroy the economically costless surplus value created for the owner and the rest of the community when personal assets are owned personally.[4]

A going concern

A country estate is a living thing. Its different parts support each other. Even in economic terms, the amputation of a single limb may cause the death of the whole. If part of the estate survives, it is less than the part of the whole that it was before.

Examples abound. Craftsmen can be used to keep up the fabric of a historic house as well as farm buildings. Forestry and market garden activities can be combined in the cultivation

---

[1] The Trust's sympathetic treatment of former owners was at the centre of a dispute among the membership in the 1960s. The objectors were defeated by a large majority at an extraordinary general meeting and have not been heard of since.

[2] *Planning or Privilege: The Case for Public Ownership of Agricultural Land* (undated), p. 12.

[3] *TUC Economic Review*, 1977; reaffirmed in Memorandum of Evidence to the Northfield Committee, *op. cit.*, para. 16.

[4] The TUC/NUAAW/Labour Party proposal is structurally similar to the present system of permitting family portraits and the like, alienated to meet tax liabilities, to be returned to the ancestral house on permanent loan. The only identifiable result of these devious manoeuvres is to destroy the surplus value created by the personal ownership of personal assets.

of trees and shrubs for sale. Own timber can be used for a variety of estate purposes. The same vehicles and machinery can serve a number of different estate departments and can themselves be serviced in a common workshop. At Milton, near Peterborough, for example, where the appearance of the estate is taken seriously, the relatively small woodlands (some 1,000 acres) have in recent years been supported by the rest of the estate to the sum of £10,000 a year and more to offset drought, gales and elm disease; farming activity also pays for the upkeep of fences, hedges, banks, verges and ditches and for not grubbing-up hedges when it would be more economic to do so. At Chatsworth, which is typical in this respect of many large houses, estate income is required for the upkeep of the house in addition to receipts from visitors. Again, the game department runs at a loss but the gamekeepers provide a service of traffic control for visitors, which is another example of complementary activities supporting each other. Large gardens are supported by the rest of the estate almost everywhere.[1]

A living whole is required, not merely for economic reasons, but to preserve the essence and spirit of a house and its estate. The National Trust recognises this: 'The Trust regards the essential unity of house, contents, garden and park and/or surrounding land as of paramount importance.'[2]

It is on the intangible value of a living community that the gap between the country and the uncomprehending townsman is widest and the recognition of the landlord's contribution most grudging.[3] For a village, as for an estate, personal owner-

[1] Although the need for a large house to be supported by an estate is now increasingly recognised, it was ignored by the Gowers Report of 1950 which had much influence on official and political thinking during the subsequent two decades. (*Report of the Treasury Committee on Houses of Outstanding Historic or Architectural Interest*; Chairman, Sir Ernest Gowers.)

[2] *The National Trust and the Country House*, op. cit., p. 2. Not only are large country-house collections the best and cheapest way of spreading fine pictures around the country; the collection and the house also enhance each other.

[3] Thus Iona was fortunate with one reporter who visited it at the time of the Argyll sale. 'The appeal of Iona', reported R. H. Greenfield in the *Sunday Telegraph* (15 April 1979), 'lies in its fragile blend of religious shrine, historic site and living community. John Black, a young crofter farming 20 acres of the island, said: "The Duke has been a good landlord. He has avoided commercialisation, but allowed us to carry out all necessary development. It is very unfortunate that he is having to sell." Angus Johnston, proprietor of the Columba Hotel and secretary of the Community Council, said: "News of the sale came as a great shock. We do not want the island exploited but neither do we want to become exhibits in a museum. We are a live community. The Duke struck just the right balance".'

ship may provide the right type of addition and enhancement, the right balance between petrification and loss of continuity and character. The existence of personally owned estates may not cure all the ills described in the report on the decline of the British village published in April 1980 by the National Council for Voluntary Organisations; but it is at least a start. At Woburn, for example, the Bedford estates are trying to maintain a mixture of shops and to prevent an excessive concentration on trading in antiques and the like. 'Perhaps the most authentic villages in scale and size', says Cormack, citing the example of Rockingham, 'are those which are owned by single families.'[1]

These values too have their external counterparts. The same argument holds good for outsiders, at least for those to whom the countryside is more than 'recreational facilities' or 'areas of outstanding scenic beauty'. It would be difficult to imagine anything further removed from the spirit of Wordsworth than the prose of the Countryside Commission; personal landowners do not talk in these terms. In addition to the saving of government expenditure, personal landownership thus serves the preferences of those who like their countryside unorganised, un-nature-trailed and generally left in peace.[2] Similar arguments apply to forests and the Forestry Commission.[3]

Unvisited, unblest?

Forests illustrate the principle of pure conservation,[4] namely, the conservation of something beautiful or noteworthy, not so

---

[1] P. Cormack, *op. cit.*, p. 71.

[2] It is probably not an accident that the signs of the National Trust are so much better designed and more discreet than those of the Countryside Commission and the Forestry Commission. This reflects the similarity of attitude between the Trust and the personal landowners. An intrusion into the countryside with no obvious purpose except to create work for the Scottish Countryside Commission was the blazing of the trail from Portpatrick to Cockburnspath. A more recent example is the Countryside Commission's proposed 'Cambrian Way' from Cardiff to Conway; and an example of a different kind is the plan to turn Shotover Hill, Oxford, into a 'country park' and thus destroy its attractions for those who like it as it is.

[3] Robert Miller, *State Forestry for the Axe, op. cit.*

[4] This is a variant of the fifth of the five principles governing the public interest in landholding systems (page 68). The examples cited there indicate that this principle is attested by people spending their own money. Nothing in this argument is intended to disparage the use of forests for ordinary commercial purposes, which is at present inhibited by capital transfer tax.

that it may be visited, used and admired, but simply so that it may continue to exist. Because of their size, forests cannot be visited easily like a house or known like a book. And yet people can and do take an interest in forests they are never likely to see, just as they may feel for the preservation of wildlife in remote regions. The mere knowledge of the existence of these things may cause pleasure and of their destruction pain.

The opposite opinion is an important influence on public life and policy. Even among conservationists, the public interest in heritage assets is often thought to be access: houses and land should be open to the public for as long as possible, and the more they are visited, the more their value to the public. The thought was more delicately expressed by Gray in his *Elegy Written in a Country Churchyard*:

> Full many a flower is born to blush unseen,
> And waste its sweetness on the desert air.

Owners are accordingly put under pressure to permit public access, notably through the offer of tax concessions on this condition.

Even before visitors became a source of income for large houses, many owners made them welcome; some have done so for centuries. At present, visitors are often not merely welcome but an indispensable source of revenue. So large estates have their own reasons for granting access to the public. But this need not imply that access is the principal or only public interest in heritage assets, or that properties which cannot or do not wish to grant access should be subject to fiscal penalties.

Lonely flowers do not waste their sweetness. Millions of people attest the value of the unseen blush with that most eloquent of testimonies, their cheque book. Much, sometimes most, of the financial support for conservation obtained from the public by voluntary organisations comes from people who never expect to see the object of their concern, whether churches or cheetahs.[1]

---

[1] This is the fifth principle governing the public interest in landholding systems (page 68). It must hold good for the supporters of a body like the Historic Churches Preservation Trust which made grants to more than 170 churches in all parts of the country during the year ended 30 September 1980, many of them small and in out-of-the-way places. Nor is the motive of the supporters primarily religious. The Trust believes that most of its financial support comes, not from 'committed Christians', but from 'the outer ring'. 'More people like

[*Contd. on p. 98*]

The same holds good locally. A house and its grounds can be an asset to the community even if always used privately. Neighbours can and do regret changes in the use of a property to which they were never invited. Heritage assets are not just famous houses and beauty spots but anything good that enhances its neighbourhood. Small houses of quality can be almost as much of a struggle to maintain for people of modest means as large houses for famous landowners. And their loss can be of the same character as the loss of a great home, even if there was never any question of opening to the public. Only the scale is different. The tax system inflicts additional damage in a different place by denying tax reliefs to owners for whom the grant of public access may be impossible, intolerable or pointless.

Conclusion

If there is a case for capital taxation at all, it does not apply to heritage assets. Capital taxation attacks the link between heritage assets and a personal owner, which is precisely what enhances their value to the rest of the community and reduces the cost and increases the quality of their upkeep.

Britain is squandering a precious and irreplaceable resource by discouraging and frustrating the historic owners of heritage assets. In economic terms, the owners' commitment to their properties is a valuable asset with no alternative use, like an expensive piece of machinery with only one purpose. In human terms, the imposition of impossible tax burdens on the owners and the continuing destruction of fine houses and the scattering of their contents cause not only anguish to their owners but also avoidable and irreversible losses to the rest of society. The taxation of heritage assets at current levels, even after the concessions contained in the Finance Act 1980, is the

[*Contd. from p. 97*]

religion to *be there* than are prepared to partake in it: more people want the Parish Church to remain standing than wish to make regular use of it. It is to this sentiment that the Trust probably makes its greatest appeal. . . . It must be to the wider field of those who value without necessarily supporting the Church that we turn in search of subscribers.' (Report and Accounts for 1979-80; emphasis in original.) Almost certainly, more than 90 per cent and probably more than 99 per cent of the financial support of the Trust comes from supporters who simply wish the churches to remain standing rather than to 'use' them for worship or even visiting. In a wholly secular context, the same argument applies to the National Trust's appeals each year on behalf of a number of different properties, and especially to its tree-planting appeal.

exception that tests the rule: this ill wind blows no good to anyone.[1]

But heritage assets do not constitute a separate philosophical category. What is interesting, beautiful or otherwise important is a matter of taste and opinion; the heritage is a seamless garment with no sharp and externally identifiable divisions. It is fortunate that there is a consensus even at the national political level on the need to prevent the destruction of the most famous properties, if necessary at the cost of reliefs from an otherwise intolerable tax régime. But this does not mean that smaller properties are less important, granted the difference in scale; merely that they inevitably have more difficulty in attracting the attention of the public. Nor is there a sharp division between heritage assets and others. The damage done by heavy capital taxes to heritage assets implies the undesirability of heavy taxes on capital and saving in general (Section VII).

---

[1] The taxation of heritage assets after the enactment of the Finance Bill 1980 is described in *Capital Taxation and the National Heritage*, HM Treasury, December 1980.

VI. THE URBAN LANDLORD

Rus in urbe

This section is shorter than the preceding one because the arguments are in general the same for the town as for the country. The main exception is that the feeling of community, present naturally in a rural estate, is now much less evident in the town.

The personally owned urban estate helps to confirm the general argument that the benefits of personal ownership to the community do not depend on public access. The public benefits from the Grosvenor estates in Mayfair and Belgravia, for example, depend scarcely at all on access but rather on the maintenance of their character as individual districts of the capital.

For a variety of reasons, personal estates account for less property in total in the town than in the country, both absolutely and relatively to all forms of property ownership. But this does not diminish the importance of the urban estate where it still exists—its importance per pound or per acre.

Rent control and 'leasehold enfranchisement'

The losses to society caused by rent control have been mentioned several times. They are essentially the same in the town as in the country. The introduction of rent controls reduces capital values overnight. And, because they interfere with voluntary economic exchange, they impose an increasing current cost year by year. From the beginning the loss falls not only on the owner but on society as a whole (owner plus sitting tenant plus potential tenants plus other third parties), and also on third parties in aggregate (society as a whole minus owner and sitting tenant). And the gains of the sitting tenant are soon exceeded by the losses of third parties, even if the interest of the owner is ignored.[1]

---

[1] The effects of rent control are discussed in several IEA publications including: Norman Macrae, *To Let?*, Hobart Paper 2, 1960; John Carmichael, *Vacant Possession*, Hobart Paper 28, 1964; F. G. Pennance, *Housing Market Analysis and Policy*, Hobart Paper 48, 1969; and especially *Verdict on Rent Control*, IEA Readings No. 7, 1972.

The income loss due to the inefficiency of rent control has its own capital equivalent. Rent control imposes an additional cost on society through the destruction of the surplus value created by the personal ownership of personal assets.[1]

Similarly for 'leasehold enfranchisement'—the compulsory transfer to the leaseholder of the freeholder's interest in leased property with compensation below market value. As with rent controls, the loss to society caused by this interference with a long-term contract between two willing parties exceeds from the beginning the gain to the beneficiary. As time passes, his gain becomes an ever smaller proportion of the loss to third parties, particularly those looking for somewhere to live. The wider loss to third parties is partly the rundown in an urban estate formerly managed as a whole when control is fragmented, for reasons discussed in the rest of this section. More generally, it is the permanent exclusion of private capital from new long-term leasehold housing development (for example, in the development of new towns) so long as the legislation or the threat of its re-introduction remains. Society could otherwise enjoy the advantages of building for letting or long lease as in the past.[2]

'Leasehold enfranchisement' is essentially an urban phenomenon in Britain and is of little or no consequence outside the towns. It illustrates the argument that concentrated benefits are more important politically than scattered losses, even if in aggregate the losses exceed the benefits. Despite the damage

---

[1] The destruction of the value of urban residential property by rent control generally amounts to some two-thirds or more: 'A sitting tenant usually decreases the value of a property by at least two-thirds.' (Sydney Chapman, *Estates Times*, 2 February 1979.) Chapman means a sitting tenant enjoying security of tenure under current legislation at a rent far below the market-clearing level.

[2] 'The remarkable thing about the London leasehold system is that it enabled the freeholder of a big estate to retain control over the use and maintenance of his property while it was on lease, and to engage in schemes of redevelopment and rehabilitation once the leases expired. . . . Only in certain towns in the British Isles, notably London, has the system of land tenure permitted the continuing centralised control of a neighbourhood that is necessary if the aims of the original plan are to be maintained, and if the plan is to be sensibly adapted to the changing needs of succeeding generations.' (Donald Olsen, *Town Planning in London—The Eighteenth and Nineteenth Centuries*, Yale University Press, 1964, Preface.)

'The concentrated holding of land would have advantages, if the owners of such land were as socially and architecturally enlightened as the great ground landlords who built up Bloomsbury, Mayfair and Belgravia in London between the seventeenth and the nineteenth centuries.' (Lewis Mumford, *City Development: Studies in Disintegration and Renewal*, Secker and Warburg, London, 1946, p. 79.)

it inflicts on society, leasehold enfranchisement is not generally thought to be reversible for years ahead; its beneficiaries are too wealthy, articulate and politically influential.[1]

The urban estate

Both personal estates and local authorities are concerned with the treatment of substantial areas of real estate as a single whole. Personal estates carried out this task long before local authorities in their modern form existed.[2] It is arguable that personal owners are more suited to this rôle than local authorities, partly because they own and manage property instead of exercising negative controls and partly because they have the same advantages of detailed knowledge, personal involvement and long-term commitment as their rural counterparts. Such advantages are denied to local politicians and officials whose personal interests are short-term, various and conflicting.

It is also certainly arguable that the present form of planning control does more harm than good and that the case for abolishing it is strong.[3] The argument extends to government *ownership* of property. In the words of the Land Decade Educational Council:

> 'Evidence from all over the world shows that personal control and involvement in housing are more socially stabilising than the collective beehive with common grounds. Indeed, it is the areas where public ownership predominates that suffer most severely from urban decay. The vision of planning control has not been as successful as was hoped, and there are many other examples where theory has not worked out in practice.'[4]

But even if government planning does do more harm than good, it does not follow that there is no rôle in planning and preservation for the large personal landowner or his institu-

---

[1] 'Enfranchisement' reduces the social value in current use of leasehold interest in an estate but increases the private value in current use.

[2] 'Most of the building development which took place on these estates in the nineteenth century was of a very high order. Because they owned large blocks of land they were able to develop them as a whole, and in the main they did so with an eye to dignity and even beauty. In London, ground landlords like the Westminsters, the Cadogans, the Bedfords and the Portmans all set a high standard of overall planning, of which the spacious squares and streets of West London are a heritage.' (Sutherland, *The Landowners, op. cit.*, p. 163.)

[3] *Government and the Land, op. cit.*

[4] *Land Decade 1780-90*, Land Decade Educational Council, London, p. 4.

tional counterparts.[1] It is not an accident, for example, that Trinity Church Square (Southwark) and its environs are an oasis of elegance and beauty in one of the drearier parts of London. The personal landowner can be more delicate in his handling of buildings than a government authority and more sensitive in his dealings with tenants and leaseholders.[2]

Planning strategy and conservation

In many parts of our cities, local government has assumed the mantle of the large personal landlord (often because the large private landowner has been taxed out of existence). 'What in effect had occurred', says Robert Thorne in his account of the move of the market out of Covent Garden, 'was that local authority planners had taken on much of the rôle once played by urban landlords, including the authority to decide about land use, densities and building renewal.'[3]

Local authorities have differed widely in the exercise of their control over property. Some have been ruthless destroyers: Camden, Hackney, Islington, Lambeth, Southwark, Tower Hamlets.[4] Others have been much more conservative, not least where the large proportion of property in private hands encouraged a policy of *laissez faire* (Kensington and Chelsea). These variations have sometimes limited the scope for co-operation between local authorities and the remaining personal landowners. But, at its best, the relationship has been harmonious and the landlord has been permitted to continue to play his traditional rôle of maintenance and adaptation with a

[1] The institutional counterparts of the large personal landowner are not the financial institutions but such bodies as religious houses, schools and Oxford and Cambridge colleges which are owners for the long term or in perpetuity and have similar attitudes and policies to those of the long-term family owner.

[2] 'It is the warmest of pleasures to renew the words I said to you last year concerning the appreciation of the Residents Association, as well as my own, for the constant care and concern you have shown for all of us.' (Letter to the Bedford Estates from the Joint Secretary of the Ridgmount Gardens Residents Association, 15 December 1975.)

[3] Robert Thorne, *Covent Garden Market—Its History and Restoration*, Architectural Press, London, 1980, p. 105. The exercise of this rôle by urban landlords is described in Donald Olsen, *Town Planning in London, op. cit.* In a comment on the high standards achieved, Olsen says: 'The building north of Great Russell Street . . . involved town planning of a most sophisticated variety, surpassed nowhere else in London.' (p. 42)

[4] For the same phenomenon in the United States, Martin Anderson, *The Federal Bulldozer: A Critical Analysis of Urban Renewal, 1949-1962*, MIT Press, Cambridge, Mass., 1964.

[103]

minimum of interference. This has the advantage of saving the local authority trouble and money and of enabling the work to be done better. The landowner and his staff know and care far more about the estate than is possible for local politicians and officials.

The most elaborate recent example of town planning by a private landowner is *The Grosvenor Estate Strategy for Mayfair and Belgravia*.[1]

> 'The maintenance of an estate in good heart is, indeed, as formidable a task as its original layout; and successive owners of the Grosvenor Estate have determined the main lines of policy, and often matters of detail as well. . . . They were certainly efficient managers of their estates, and right down to the present day there is a distinctively well-kept air about the streets and squares of the Grosvenor Estate. . . . Without good management Grosvenor Square might in the course of time have become as down-at-heel as Soho Square or Golden Square, Belgravia would never have assumed its graceful ambience, and the insertion of a railway terminus (Victoria) into the existing fabric could have had disastrous results.'[2]

The prevention of deterioration and if necessary its reversal (as in the Cadogan property of Oakley Gardens, Chelsea) are part of the business of maintenance and adaptation. And personal ownership is apparently one of the most substantial safeguards against gross errors of taste and judgement. The construction of a skyscraper on the cliff at Meads, Eastbourne, would not have been possible if the decision had rested with the Duke of Devonshire whose forbears erected the surrounding property.

In his discussion of the aristocratic urban landowners, David Cannadine writes:

> 'Almost without exception, their land was held under elaborate statute or settlement either by "families concerned to preserve and enhance their long-term social and financial position, or by corporations equally concerned about the future". Neither the desire to maximise profits, nor immediate financial concerns, was their major preoccupation: how could it be, as statutes and settlements almost invariably prohibited sale? So the long-term value of their estates and the maximisation of prestige were more

---

[1] Chapman Taylor Partners, 1971.

[2] Francis Sheppard, 'The Grosvenor Estate 1677-1977', in *History Today*, November 1977, p. 730.

important considerations. As a result, leasehold development was preferred: it gave the owners an opportunity to draw up and implement a coherent estate plan.'[1]

It is precisely this long-term personal planning that is under threat from the attack on leasehold called 'leasehold enfranchisement'. Cannadine goes on to cite examples of the long time-horizon which the estates have had in the past and would like to be able to retain:

> 'The Calthorpe Estate redevelopment plan for Edgbaston, published in 1958, confidently looked forward "to the year 2,000 A.D."; and George Ridley, the senior trustee of the Grosvenor Estates, could bravely affirm that "our ability is to look far far ahead. Fifty years is nothing to us, and one hundred years is normal".'[2]

But these worthy aims are subject to the hazards of capital taxation. The piecemeal sale of the Marquis of Northampton's Islington estate, for example, has resulted in the decay of a formerly unified whole. Again, the intrusion of the university into Bloomsbury has spoiled the domestic character of the district; even when domestic buildings have been taken over and retained, the character of the older buildings has not been respected.

Conservation and planning details

A personal estate makes its mark on a city in matters of detail as well as in the broad lines of planning strategy. Attention to detail can contribute handsomely to the harmony, elegance and character of a neighbourhood.

The Portman estate, for example, exercises control not only over the construction and extension of buildings but also over such details as paint colours, brass fittings, door lamps, railings, the preservation of trees and the materials to be used for the repair of walls. These restrictions are generally acceptable to leaseholders and tenants.

The Grosvenor standard lease for a private dwelling house includes covenants which require prior approval for the type

---

[1] David Cannadine, *Lords and Landlords: The Aristocracy and the Towns, 1774-1967*, Leicester University Press, 1980, p. 393, citing F. H. W. Sheppard, *London 1808-1870: The Infernal Wen*, Secker and Warburg, London, 1971, p. 92.

[2] *Ibid.*, p. 427, citing Calthorpe Estate Company, *Window on Edgbaston*, 1958, p. 19; O. Marriott, *The Property Boom*, Hamish Hamilton, London, 1967, new edn., Pan Books, London, 1969, p. 112.

and colour of external paints and for the manner of cleaning the stonework. They also prohibit black pointing of brickwork, painting of external parts not usually painted in the past, alteration of chimney stacks and chimney pots, and erection of television aerials without written consent. *The Grosvenor Estate Strategy for Mayfair and Belgravia* concerns itself with lamp standards, traffic signs and meters, railings, seats, tree planting, and paving materials including the patterns of cobblestones.

Local authorities have little talent for these matters and generally little interest in them either. Street lamps are an illuminating example. Street lighting is a classical public good and its provision is a classical government function. Yet the design of lamp standards is often ugly in itself and in conflict with surroundings of any quality.[1]

Leaseholders and tenants accept restrictions which serve a useful purpose and are exercised by people with a long-term commitment to a property and a feeling for its requirements. This can be provided by the personal owner and his institutional counterparts; it seems unlikely to be found elsewhere.[2]

Conclusion

The conclusion is the same as for the personal estate in the country (Section V). Confiscatory taxes on heritage assets destroy a precious and irreplaceable asset in the knowledge and dedication of the historic owners. No one else will look after the property so well or so cheaply. The losers are not only the owners but the rest of the community as well.

---

[1] *The Grosvenor Estate Strategy* cites Belgrave Square (p. 173). Another example, under the noses of our legislators, is the pair of ugly and unsympathetic lamp standards outside The Albert, one of the few buildings of any quality in the concrete-and-glass desert of Victoria Street.

[2] Tenants of local authority houses, by contrast, are justified in resisting controls on the colours they paint the exteriors of their homes when the architectural quality is such that variety does no harm and may do good.

VII. THE BURDEN OF TAXATION

The weight of the burden

The *quantitative burden* of taxation in Britain—that is, the ratio of tax revenue to national income without weighting for differences in the nature of different taxes—is not exceptionally heavy among industrialised countries. What gives Britain her well-deserved notoriety as a high-tax country is the exceptionally heavy *qualitative burden*—painful and keenly-felt taxes and high tax rates at the top of the scale.[1]

Up to the present there has been little change in these historic relationships. The tax on investment income was reduced to a maximum of 75 per cent in the June 1979 budget. But capital gains tax and capital transfer tax rise as high as ever and are among the highest in the world. Indeed, capital transfer tax has been growing heavier, since the graduated schedule of rates has not been adjusted for inflation since the tax was first enacted in 1975. In contrast with the situation in most other countries, capital taxes cannot be paid out of income over most of the range. The tax system is intended to frustrate and prevent long-term saving; and the 1975 capital transfer tax was designed to close the 'loopholes' through which saving in perpetuity had until then remained possible.

A tax on vocation

Taxes on vocation are the only unavoidable taxes, apart from poll taxes (now a historical curiosity)[2] and taxes on the neces-

---

[1] The situation described in my *The Camel's Back: An International Comparison of Tax Burdens*, Centre for Policy Studies, London, 1976, is updated in my *The Taxation of Industry: Fiscal Barriers to the Creation of Wealth*, Panopticum Press, Upminster, 1981. The facts in this book are unchanged by the March 1982 Budget, which is taken into account throughout the present section.

[2] As a real-life tax, poll taxes are still no more than a historical curiosity. But their economic merits have continued to attract interest, most recently in the Green Paper, *Alternatives to Domestic Rates* (Cmnd. 8449, HMSO, December 1981). A considerable element of opinion appears to regard a poll tax as the most attractive or least unattractive alternative to domestic rates—at least if the substitution is partial rather than total and if there can be any assurance that the extension of the tax base constituted by the addition of a poll tax will not simply lead to an increase in government expenditure and thus in the aggregate tax burden.

saries of life. Taxes on vocation have the unattractive quality of penalising work and conduct intended to benefit others. This is bad enough when a tax is levied on vocational earnings.[1] But at least taxes on vocational earnings cannot exceed 100 per cent of income. Taxes on vocational ownership, by contrast, can be a large multiple of the income from the vocational assets; they can be designed, like the British capital transfer tax, to bite deeply into the capital.[2]

The weight of capital taxes on vocational ownership is compounded by the conceptual over-valuation for capital tax purposes of assets in general and heritage assets in particular. Assets in general are systematically over-valued for purposes of capital transfer tax because the market values on which they are based are largely determined by taxpayers subject to the tax at lower rates or not at all; for them the assets are correspondingly more valuable than for taxpayers subject to higher rates of the tax. Heritage assets are systematically over-valued because the tax is based on break-up value whereas their worth to the taxpayer in his private capacity—the private value in current use—may be much lower. This conceptual over-valuation of heritage assets may far outweigh the reliefs granted to the assets from the full burden of the tax.[3]

Two senses of 'taxable capacity'

Graduated or 'progressive' taxation (taxation at proportionately higher rates as income or capital rises) is sometimes justified on the ground that it charges in accordance with 'ability to pay' or 'taxable capacity'. As between richer and poorer, it is a form of charging what the traffic will bear. But there is a second sense of 'taxable capacity' which distinguishes not between richer and poorer taxpayers but between *necessaries* and *luxuries*—goods bought relatively more by poorer

---

[1] *Vocational earnings* are earnings that are insufficiently remunerated to compensate for the time and trouble spent on an activity which would have diminished or ceased but for the vocation. (*Tax Avoidance and Evasion, op. cit.*, p. 94.) People will work for nothing in vocational causes and even at their own expense. For the relationship between the taxation of vocational earning and the taxation of vocational ownership, above, page 83, note 2.

[2] For the income equivalents of taxes on capital, *Is Capital Taxation Fair?, op cit.*, Ch. IV, especially pp. 91-103.

[3] Legislation on heritage taxation up to and including the Finance Act 1980 is described in the Treasury's Memorandum, 'Capital Taxation and the National Heritage', December 1980.

and richer respectively. Charging what the traffic will bear implies that necessaries should be taxed proportionately more than luxuries since the demand for necessaries is less sensitive to variations in price.[1] These two senses of 'taxable capacity' thus conflict in that charging what the traffic will bear favours the poor relatively to the rich in the graduation of tax schedules and *vice-versa* in the distinction between luxuries and necessaries.

Ownership is in the technical sense a luxury (it becomes relatively more important as income rises), although the ownership of heritage assets is in human terms not so much a luxury as an expensive and anxious responsibility. Luxuries in the technical sense are a poor tax base because taxpayers can do without them and so taxes on them can easily be avoided. This effect is re-inforced when break-up value is far in excess of private value in current use; the only motive then inducing the owner of heritage assets to carry on is vocation, and vocation alone may not be enough. The inescapable burden of running a large estate in conditions for which it was not designed is gratuitously and doubly aggravated by the tax system: first, because taxes on large estates are heavy in general, and, secondly, because they are particularly heavy when the estate consists primarily of heritage assets since personal far exceeds private value in current use and the tax system is at odds with the benefit to society obtainable from vocational ownership. Often the tax system wins by brute force; the tax charge levied on death leaves the owner's successors no option but to dismember the estate. Sometimes fiscal inevitability is anticipated by a collapse of morale. The owner who knows that, short of a change in the tax régime, no effort on his part can do more than delay the dismemberment of his estate (a process from which the rest of the community may well lose more than he himself and his family) is likely to feel with Lady Monchensey:

> 'I will let the walls crumble. Why should I worry
> To keep the tiles on the roof, combat the endless weather,
> Resist the wind? Fight with increasing taxes
> And unpaid rents and tithes?'[2]

[1] The income elasticity of demand (the responsiveness of demand to a change in income) is less than unity for necessaries and more than unity for luxuries.
[2] T. S. Eliot, *The Family Reunion*, Part II, Scene III.

The taxation of saving in perpetuity

Taxes on investment income and its parent capital are different forms of tax on saving. Since the value of saving is what it will buy, there is no need in a consistent tax system for any form of tax on saving in order to achieve neutrality between saving and spending; taxes on spending are always and everywhere sufficient to achieve that neutrality.[1] Taxation of investment income or its parent capital constitutes a fiscal bias against investment and thus causes a loss to society through the diversion of resources from saving to spending. Nor is any form of tax on saving required to assist a policy for the egalitarian redistribution of spending. On the contrary, the inequality of spending is increased, not reduced, by high tax rates on saving since they reduce the opportunity cost of spending by the richer taxpayer.[2]

Of the industrialised countries, only Australia levies no tax on capital—either transfers or gains. There is no federal tax on capital transfers in Canada (and no provincial tax except in Quebec). All the other industrialised countries have some form of tax on capital transfers (which implies heavier taxation of long-term saving than of short-term). But in so far as tax is levied in the other countries on capital gains, the distinction is invariably in the opposite sense—gains in the short term are taxed more heavily than in the long. Official policy in most countries thus vacillates incoherently between a preference for saving in the long term and in the short.[3]

Despite minor qualifications, the British tax system possesses a heavy bias against long-term saving relatively to short-term. Capital transfer tax, which affects only long-term saving, is among the most onerous in the world; and capital gains tax is not reduced, as in a number of other countries, on assets held for more than a stated period. The loss to society, as well as to the taxpayer, is most acute in land because it is held for the long term; in a neutral tax environment, land is often handed

---

[1] *Is Capital Taxation Fair?*, op. cit., pp. 59, 89.

[2] *Ibid.*, pp. 61-67.

[3] Similarly, British tax relief on life assurance premiums favours long-term saving over short-term within the lifetime of the taxpayer. This favour is reversed when the taxpayer dies because long-term saving incurs a liability to capital transfer tax which it would have escaped if the saving had been realised and spent. The argument for capital gains tax is essentially in favour of assimilating short-term saving to short-term trading. It disappears when the saving is in perpetuity and gains merely represent a re-arrangement of portfolios. (*Ibid.*, pp. 93-103.)

down for generations or centuries, even within families of modest means. The tax system is stacked against land ownership because of its long-term character, and present reliefs are at best a partial palliative.

The bias is illustrated most vividly by the farmers' paradox (above, page 28). The rationale of 'progressive' taxation is to levy on taxable capacity. Yet there is no taxable capacity in the capital value of an asset held in perpetuity, and the attempt to tax it makes the taxpayer poorer when the value of his assets rises. Taxes on transfer at death are especially damaging because of their capricious and unpredictable incidence.

A strong case can be made for not taxing saving at all and levying exclusively on expenditure.[1] It is especially strong when the saving represents a long-term commitment and not a source of liquid funds. When land and other heritage assets are subjected to taxes which cannot be paid out of income, the losers are not only the owners but the rest of the community as well. And the public may lose much more than the owners.[2] This is a particular illustration of the general argument that the cost of the 'progressive' taxation of income and capital falls largely or even primarily on third parties.[3]

Tax alleviation or tax reduction?

The tax burden on heritage assets is a problem common to a number of industrialised countries apart from Britain. But it is aggravated in Britain by the exceptionally heavy structure of capital taxation. If it is accepted that to subject heritage assets[4] to the full rigours of a high-tax system would be intoleraby destructive, there is a choice to be made.

[1] *Ibid.*, Ch. IV. Also my *A Liberal Tax Policy, op. cit.* The rationale of expenditure taxation, whether proportional or graduated, leaves no scope for taxes on capital. I have argued that this is a central flaw in the analysis of the Meade Report on *The Structure and Reform of Direct Taxation* for the Institute for Fiscal Studies. (My 'The Meade Report and the Taxation of Capital', *British Tax Review*, 1/1979.)

[2] The excess of social value in current use over personal value in current use may be a substantial multiple of the excess of personal value in current use over break-up value.

[3] In particular because the 'progressive' taxation of income and capital reduces the opportunity cost of spending at the top of the scale. The argument is set out in my *The Poor Pay Most: The Paradox of 'Progressive' Taxation*, Bow Group, London, 1979.

[4] The argument is the same for 'productive assets', such as farmland and unquoted companies, which represent sunk capital and not a potential source of liquid funds for spending or the payment of taxes.

It is possible to maintain the *form* of the high-tax system but to introduce concessions specifically designed to mitigate the damage to heritage assets. The argument against this option is that heritage assets are not a distinct category and that whether an asset should qualify for such reliefs is not a suitable question for the administrative discretion of Treasury or Revenue officials, or indeed anyone else.[1] Reliefs where the shoe pinches worst are a pragmatic means of reducing the damage done by taxation. But they still leave the full rates of tax payable by other taxpayers who may think they have an equally good case for relief. The maintenance of a high-tax régime rendered less painful by tailor-made concessions to accommodate particular lobbies and interest-groups is the opposite of the Thatcher Government's policy for income tax, which is to cut both the rates and the reliefs (the reduction or withdrawal of reliefs being additional to the reduction in their value resulting from the cuts in tax rates).

The other option is to reduce the tax rates themselves and thus extend the reductions to all taxpayers affected. This costs more in tax revenue forgone; but it avoids the economic distortions created by all high tax rates and the unfairness of levying them only on taxpayers not organised into interest-groups sufficiently effective to obtain reliefs.

What has happened in this country since the war is that a high-tax régime has been maintained, alleviated by reliefs[2] for heritage assets too late and too grudging to prevent the destruc-

---

[1] An additional consideration is that good intentions are likely to be strangled in red tape: 'A number of legal advisers, trustees and owners of works were disinclined to seek exemption because of "administrative complications and pressures".' (Hugh Leggatt, Secretary of Heritage in Danger, reported in the *Daily Telegraph*, 20 January 1981.)

[2] Tax reliefs are the principal offset to the high-tax structure. They are occasionally supplemented by emergency grants, such as the £1 million-plus to Kedleston Hall, Derbyshire, in 1980 from the National Heritage Memorial Fund, in addition to the grants available for current costs. Patrick Phillips of Kentwell Hall, Long Melford, does not apply for grants towards the rising cost of upkeep and repairs, not only on grounds of principle but also for practical reasons: 'The Government makes the rules, so that you often have to get a much more expensive job done than you would otherwise have considered necessary. And, since the grant only covers a part of the cost, I am not convinced that at the end of the day you really save anything at all.' (Quoted in John Young, *The Country House in the 1980s*, Allen and Unwin, 1981, p. 103.) This illustrates the general argument that tax reliefs are superior to government expenditure as a means of providing funds for the conservation of heritage assets (as indeed they are for most other economic purposes).

tion of buildings and the dispersal of their contents on a scale not experienced since the dissolution of the monasteries. The contention of this *Hobart Paper* is that there is no sharp distinction between heritage and other assets and that the arguments for reducing—and preferably abolishing—capital taxes on heritage assets are also valid, though less strong, for land and other assets in general.

Recommendations for public policy

The simplest and most attractive policy option is to abolish the taxes on capital gains and capital transfers. In Australia, for example—a country with which Britain has much in common socially and politically—there has never been a tax on capital gains, and federal death duties and gift tax have been abolished in the last few years. Capital gains tax is mostly a tax on inflation and capital transfer tax falls largely on productive assets and the heritage; if these elements were removed from the tax base, the taxes would hardly be worth collecting.[1] The abolition of the capital taxes could form part of a move towards a more neutral system in which the ownership of land was not disproportionately burdened with taxes or favoured with subsidies and in which taxation and government expenditure were both reduced.[2]

If capital taxes are not abolished, they should at least be reduced to such rates as can be paid out of *income* by any taxpayer. This conclusion holds good not only for heritage and

---

[1] The case for abolishing capital transfer tax in particular and capital taxes in general is argued in my *A Tax for the Axe: The Case for Abolishing Tax on Capital Transfers*, CUT, the Taxpayers' Union, London, January 1980.

[2] An extreme example of the double counting and destruction of surplus value in the present taxation of heritage assets is the practice whereby the value (or 'net cost') of assets accepted by the Inland Revenue in lieu of capital transfer tax is treated as expenditure half by the Department of the Environment and half by the Office of Arts and Libraries (these two bodies having succeeded the National Land Fund in this capacity in 1980). All the parties lose from this arrangement (criticised by the House of Commons Select Committee on Education, Science and the Arts in April 1981), either relatively or absolutely or both. The taxpayer loses because he has to part with heritage assets he would prefer to keep. The Revenue loses because it receives less money than if tax were levied without provision for 'in lieu acceptance' (it loses at present by not less than 25 per cent of the tax otherwise payable). The Department of the Environment and the Office of Arts and Libraries lose by having items imposed on their votes which are only of notional or book-keeping significance and need not be there if the assets remained in the hands of the taxpayer. The Treasury loses because its policy is to *reduce* taxation and government expenditure and the present system *increases* both.

productive assets but for assets in general. The case for special treatment of heritage assets is that capital taxes should be levied, if at all, on something near the private value in current use and not on the much higher break-up value.

Tax concessions on heritage assets may raise rather than reduce the yield of tax revenue. Incentives in the US Tax Reform Act, which allow owners to write off against tax the cost of rehabilitation for five years, had not cost the US Treasury more than $10-20 million in any one year; yet, by January 1980, they had already resulted in $640 million being spent on conservation during the Act's life of less than four years.[1]

Tax concessions on heritage assets should not be conditional on public access. Failing that, the condition should be satisfied either by occasional opening to the general public or by the granting of access by appointment to particular visitors or groups.[2]

The levying of value added tax on *repairs and maintenance*, though not on *demolition or new construction*, is an invitation to destroy old buildings which ought to be repaired and restored. It is a fiscal attack on the idea of conservation, as much case law shows. Value added tax on repairs has no serious justification within the present tax system. Preferably it should be abolished at once; if not, it should be phased out over a period of a few years.

Reforms along these lines would reduce the need for the expenditure of government money on heritage assets. Personal ownership is the more economical alternative.

---

[1] *Britain's Historic Buildings: A Policy for their Future Use*, British Tourist Authority, 1980. If these figures are correct, they imply a gain from the incentives concerned not only for the US economy but even for the US Internal Revenue Service; it is difficult to interpret the figures in any other sense.

[2] The argument applies to taxes on income as well as on capital. At present, running costs can be set against income tax only if a house is opened to the public as a serious business venture intended to make a profit. Otherwise only administrative, and not maintenance, costs are allowable. Viewing by appointment is already one of the ways in which the condition of 'reasonable public access' may be satisfied in order to secure exemption of works of art from capital transfer tax and/or capital gains tax. (Treasury Press Release, 'Conditional Tax Exemption for Works of Art', 14 January 1981.)

VIII. CONCLUSIONS

This *Paper* has argued that personal ownership of land and heritage assets increases economic welfare because the satisfaction or utility obtainable from ownership is additional to the utility from consumption. Land and heritage assets are *personal* and the additional utility is diminished or destroyed if they are owned impersonally, especially if they are owned by the state or its agencies.

The public interest in the personal ownership of land and heritage assets is twofold. Third parties generally gain (the external effects are positive) because the economic costs of management and custody are subsidised by the pleasure of ownership; a higher standard of management and custody is obtainable more cheaply. But the personal owner is also a member of the public, and his interest is part of the public interest.

The argument is general; there are no sharp divisions between land and other assets, between heritage assets and others, or between town and country. Any differences are of degree and not of kind. While the proprietary principle of wealth creation applies most strongly to personal assets, it also applies to the much wider category of assets that are proprietary without being personal. The sale of the National Freight Company to its employees is a case in point. There are few assets to which the principle does not apply. Thus it lends weight to the case for the privatisation of both nationalised industries and social services (health, education, pensions, etc.) in addition to the more familiar argument concerning increased efficiency on income account. The blight of sterile ownership can afflict proprietary assets which are impersonal as well as those which are personal.

In this context, taxation is the single most damaging form of government interference in the economy, since it harms not only personal assets in particular but proprietary assets in general. This *Paper* argues that capital taxes are always disproportionately destructive because they strike directly at the economic satisfaction obtainable from ownership. Taxes on investment income, however, strike at capital indirectly by reducing the value of its yield. The only taxes immune from this criticism are those on spending and the basic rate of income tax on earnings; a proportional tax on earnings is substantially equivalent to a proportional tax on spending for the large number of taxpayers who spend the whole or almost the whole of their net-of-tax income. The creation of wealth through ownership is all the more severely damaged if the taxes on capital, investment income, or even earned income are not proportional but graduated.

QUESTIONS FOR DISCUSSION

1. What advantages does the ownership of assets confer on a personal owner in addition to the possession of a capitalised income stream?

2. What is the economic rationale of saving in perpetuity?

3. Which assets are typically personal, and which assets are proprietary ones without also being personal?

4. Is land more proprietary and more personal than other assets?

5. How could the capital-efficiency of land ownership be increased?

6. Are there advantages to third parties from the ownership of rural or urban land by the state rather than by personal owners?

7. What forms of occupation and ownership are typically vocational?

8. What classes of heritage asset, if any, are more suitable for ownership by the state than by personal owners?

9. Should tax reliefs on heritage assets be made conditional on public access?

10. What are the policy implications of maximising the welfare obtainable from personal ownership?

FURTHER READING

Anderson, Martin, *The Federal Bulldozer: A Critical Analysis of Urban Renewal, 1949-1962*, MIT Press, Cambridge, Mass., 1964.

Bracewell-Milnes, Barry, *Is Capital Taxation Fair? The Tradition and the Truth*, Institute of Directors, London, 1974.

——, *The Taxation of Industry: Fiscal Barriers to the Creation of Wealth*, Panopticum Press, Upminster, 1981, through The Alternative Bookshop, Covent Garden, London, WC2.

Cheung, Steven, *The Myth of Social Cost*, Hobart Paper 82, IEA, 1978.

Cormack, Patrick, *Heritage in Danger*, New English Library, London, 1976.

Cornforth, John, *Country Houses in Britain—Can They Survive?*, Country Life, London, 1974.

Denman, D. R., *The Place of Property*, Geographical Publications, 1978.

Furubotn, Eirik G., and Pejovich, Svetozar, *The Economics of Property Rights*, Ballinger Publishing Company, Cambridge, Mass., 1974.

MacCallum, Spencer Heath, *The Art of Community*, Institute for Humane Studies Inc., Menlo Park, California, 1970.

Northfield, The Rt. Hon. Lord (Chairman), *Report of the Committee of Inquiry into the Acquisition and Occupancy of Agricultural Land*, Cmnd. 7599, HMSO, 1979.

Warriner, Doreen, *Land Reform in Principle and Practice*, Clarendon Press, Oxford, 1969.

New Titles from the IEA

Occasional Paper 62
Could Do Better
Contrasting assessments of the economic progress and prospects of the Thatcher Government at mid-term
MICHAEL BEENSTOCK, JO GRIMOND, RICHARD LAYARD, G. W. MAYNARD, PATRICK MINFORD, E. VICTOR MORGAN, M. H. PESTON, HAROLD ROSE, RICHARD C. STAPLETON, THOMAS WILSON, GEOFFREY WOOD
March 1982   112pp   £2·80

'The Institute of Economic Affairs's latest *Occasional Paper* provides the comments of 11 good men and true whose opinions range from the complete confidence in an economic miracle of Prof. Richard Stapleton of Manchester Business School to the completely opposed views of his opposite number at the City University Business School.'   Pritchard, *Yorkshire Post*

'Far from being hagiographical, the report comes from a wide variety of hands, some of which, such as those of Mr Jo Grimond or Prof. Richard Layard, bear the Government little good will. But it is remarkable how much agreement there is both on the historical outlines of the past two and a half years, and on the needs for the future. Most contributors criticise the Government in its early stages for a failure to tackle the trade unions, and a reluctance, expressed partly in high public sector pay awards, to control public spending. They say the Government has been unadventurous, and that it has tended to believe that monetary policy is a substitute for institutional reform.'   Leader, *Daily Telegraph*

Hobart Paper 92
What Price Unemployment?
An Alternative Approach
ROBERT MILLER and JOHN B. WOOD
January 1982   80pp   £1·80

'The deputy director of the Institute, Mr John B. Wood, and a researcher, Mr Robert Miller, say that, contrary to popular mythology, the reduction of unemployment depends to a considerable degree on spending cuts.

'Government spending is a potent cause of unemployment, since taxing the lowest paid becomes necessary to finance it, with devastating consequences on incentives to work.

'The authors argue that if the Government wishes to make a permanent reduction in the natural rate of unemployment, it must attempt a radical reform of the labour and housing markets.'
Robert Scott, *Yorkshire Post*

Hobart Paperback 14
The Emerging Consensus
Essays on the interplay between ideas, interests and circumstances in the first 25 years of the IEA
Edited by ARTHUR SELDON; authors include GRAHAM HUTTON, T. W. HUTCHISON, J. M. BUCHANAN, DAVID COLLARD, LORD CROHAM and COLIN CLARK
June 1981   xxxv+284pp.   Paperback £3·60, Cased £6·00

Hobart Paper 90
How to End the 'Monetarist' Controversy
SAMUEL BRITTAN
July 1981   132pp.   £2·50
Subtitled 'A journalist's reflections on output, jobs, prices and money', this book by the principal economic commentator of *The Financial Times*, provides a far-ranging analysis of current economic policy issues facing the UK.

'... the latest paper from the indispensable Institute of Economic Affairs ... [goes] back to first principles, and by stripping away the political grime maximises the area of consensus between so-called monetarists and so-called Keynesians.'
Leader, *Daily Telegraph*

'Mr Sam Brittan of the *Financial Times*, a doyen of British economic journalists ... wants the British treasury to choose as its annual target a steady rise in money gdp. This would be a return to original Keynesianism, and a logical next step in Britain's past 50 years of running round in macro-economic circles ...'

'It is quite probable that Mr Brittan's [proposal] would be better than most other stopping places round the ring in giving the authorities some sort of quantitative grip on what they think they are doing.'
Economist

Occasional Paper 60
Wither the Welfare State
ARTHUR SELDON
September 1981   48pp.   £1·50
'Recent developments seem to be lending support to a thesis which Arthur Seldon has been propounding for many years, namely that the British welfare state will meet increasing strains in the attempt to perpetuate itself in the face of market forces which conflict with it.'
Banker

Research Monograph 36
Manufacturing Two Nations
The 'sociological trap' created by the bias of British regional policy against service industry
JOHN McENERY
October 1981   48pp.   £1·50
'Mr McEnery ... argues persuasively that regional policy has been working against the grain.'   Leader, *Financial Times*
'An important new study.'   *Morning Telegraph* (Sheffield)
'Mr McEnery's analysis hits plenty of nails bang on the head.'
Financial Weekly

Occasional Paper 61
The Disorder in World Money:
From Bretton Woods to SDRs
Twelfth Wincott Memorial Lecture
PAUL BAREAU
December 1981   32pp.   £1·00
A distinguished financial journalist, Paul Bareau discusses 'the decline of monetary morality over the past 35 years' and argues for the reform of the international currency system.